KU-134-013

Artists' Pigments c. 1600–1835

A Study in English Documentary Sources

Technical Studies in the Arts, Archaeology and Architecture

Series Editor

Derek Linstrum

Institute of Advanced Architectural Studies,
University of York

Technical Studies in the Arts, Archaeology and Architecture

Artists' Pigments c. 1600–1835

A Study in English Documentary Sources

Second Edition

R. D. Harley

Butterworth Scientific

London Boston Durban Singapore Sydney Toronto Wellington

First published 1970
Second edition 1982

British Library Cataloguing in Publication Data

Harley, R.D.
Artists' pigments c. 1600–1835.—2nd ed.—
(Technical studies in the arts, archaeology and architecture)
1. Painting, English—Bibliography
2. Pigments—Bibliography
I. Title
759.2'07 ND466

ISBN 0–408–70945–6

Typeset by Scribe Design, Gillingham, Kent
Originated by CJS Photolitho Services Ltd
Printed and bound in Great Britain by Butler & Tanner Ltd, Frome

Foreword

Rutherford J. Gettens
Formerly President of the International Institute for Conservation, and Research Consultant, Technical Laboratory, Freer Gallery of Art

Historians of art, curators of paintings and museum scientists all over the world are showing increasing interest in the materials that paintings were made of and how those materials were put together to make a finished work. Much time is spent identifying pigments and mediums by all kinds of modern methods of chemical and physical analysis, and examining cross-sections and X-rays to learn about the layered structure of paint films. In addition to skill in analysis, the technical researcher needs to know the historical background of painting techniques and to be acquainted with the technology of materials going back many centuries. This in part has been supplied by translators of Theophrastus, Pliny, Vitruvius, Theophilus, and a long list of medieval manuscripts, which include such names as Sir Charles Eastlake, Mrs Mary P. Merrifield, E. Berger, Daniel V. Thompson, J. A. van de Graaf, V. and R. Borradaile and others. These people have produced notable reference books invaluable to modern investigators in all aspects of painting.

Dr Harley, after thorough and painstaking studies, will join them by filling one of the big gaps, namely the history of materials of English painting, here revealed in documentary sources little known and almost unavailable to the average investigator. The period covered is important because it leads into the era of Reynolds, Gainsborough, Turner and the great English school of water colourists.

R.J.G

Preface

to the Second Edition

In the ten-year interval between publication of the first edition and preparation of the second, work on the history and identification of pigments has increased and I have been able to benefit from the interest and expertise of others.

I should especially like to thank Miss Joyce Plesters and her colleagues at the National Gallery whose generosity in discussing their work on the identification of pigments has provided such useful complementary material to this edition. Over the past ten years it has been a constant source of surprise to me that some who have read the book gain the impression that I am a chemist, when in fact my knowledge is the product of training as an historian and artists' colourman. I can only think that the benefit of having had enjoyable discussions with scientists, such as Miss Plesters, has added something to the book, apart from giving me added confidence that, more often than not, documentary sources have given a reliable commentary on the history of use of artists' colours.

It must be understood that wherever I state that the earliest reference to any colour is to be found in a certain source or that no reference has been found, I am referring to the particular sources I studied, as outlined in Chapters One to Three. It is not only possible but likely that certain views will come to be modified as new documentary evidence is studied. The continuation of research is reflected in references in this edition to several university theses and to the papers in *Studies in Conservation* in the 'Identification of the Materials of Paintings' series.

Continuation of my own research has led to inclusion of material found in George Field's manuscript notebooks that I knew had survived into the early twentieth century but had been unable to find while undertaking my postgraduate research; they have now been placed on loan at the Courtauld Institute of Art, University of London. They form an important source for the history of artists' colours in the early nineteenth century that is comparable with the importance of two of the Mayerne manuscripts of the early seventeenth century. Some new information has also been derived from the eighteenth- and nineteenth-century records of Berger, Jenson & Nicholson Ltd. I have to thank Mrs Kate Lowry and Ian Bristow for independently drawing my attention to Berger's archives, and Ian Bristow especially for permitting me to consult the first part of his postgraduate thesis prior to its presentation for the degree of D.Phil.

Finally, I should like to thank the International Centre for the Study of the Preservation and the Restoration of Cultural Property, Rome, for the contribution that has made possible the inclusion of illustrations, including colour plates.

R. D. H.

Preface

to the First Edition

This work is a slightly revised version of my thesis 'Documentary Sources on the History of Artists' Colours in England, *c.* 1600–1835', which was presented in 1967 for the University of London degree of Doctor of Philosophy in History of Art. I would like to thank Mr S. Rees Jones of the Courtauld Institute of Art for much helpful encouragement throughout my time as a research student and subsequently, and also the Board of Directors of Messrs Winsor & Newton Ltd for allowing me time to undertake the research.

I wish to acknowledge the encouragement and assistance of several members of the International Institute for Conservation, especially that given by Dr A. E. Werner, who read the text and suggested several improvements, and by Mr G. Thomson, who gave much help towards publication.

Amongst others who took an interest in my work, I would like to thank the Paul Mellon Foundation for British Art. The staff of the University of London Library were particularly helpful and, in conjunction with the National Central Library, succeeded in tracing several rare books in Britain and America, and made arrangements for photography.

As the title indicates, original research was concentrated on English documentary sources, but reference is also made to other literature relating to pigments and painting in Europe. Where possible, generally accepted chemical formulae and botanical names have been included; most have been taken from two standard works: *Painting Materials* by R. J. Gettens and G. L. Stout, and *Colour Index* published by the Society of Dyers and Colourists.

R. D. H.

Acknowledgements

The Publishers gratefully acknowledge the permission of the following to reproduce illustrations.

Ann Ronan Science Library, Figs 19, 28, 30; Arthur Ackermann & Son Ltd, Fig. 54; British Library, Fig. 18; the Trustees of the British Museum, Figs. 1, 5, 6, 8, 9, 40 and Plates 7, 8; Courtauld Institute of Art, Figs. 27, 35, 49, 52, 53 and Plates 1–6; National Portrait Gallery, Figs. 3, 12, 24; National Swedish Art Museums, Fig. 50; Royal Horticultural Society, Figs. 20, 22, 32, 33, 42; Science Reference Library, Fig. 17; University of London Library, Figs. 13, 14, 15, 38, 39, 45, 47; Victoria and Albert Museum (Crown copyright), Figs. 2, 4, 7, 29, 46; Messrs Winsor and Newton Limited, Figs. 10, 16, 21, 25, 31, 36, 37, 41, 43, 44, 51. Fig. 34 is reproduced from the Gorhambury collection by permission of The Earl of Verulam.

Contents

1

Literary Sources: Manuscripts and Printed Books, Sixteenth Century to Mid-Seventeenth Century

'Too many colours already' is the retort attributed to Richard Wilson upon being told of a new pigment. The painter doubtless advocated limiting the palette to a few established artists' colours, and yet his dismissal of a new pigment may surprise those who are aware of the number of useful pigments which were introduced during the eighteenth and nineteenth centuries and which have remained as important artists' colours until the present time. However, a further study of the transition period during which a number of pigments used in medieval times fell into obsolescence and were replaced by modern pigments suggests that eighteenth-century artists such as Wilson may have had good reason to regard innovations with caution, for not all new pigments were satisfactory. Some which are now accepted were not as well made then as they are now, whereas others were soon discarded as being fundamentally unsuitable for artists' use. The history of pigments is intimately connected with the development of technology and chemistry, yet it is incomplete without a study of artists' views on the subject. Manuscripts and books on painting frequently provide valuable information, especially on the early stages of the transition from a medieval to a modern range of pigments.

There is fortunately no lack of material concerning the majority of traditional colours. For the period before 1600 one must necessarily rely very much upon manuscript material, although two books on painting were printed in English before that date. Manuscript sources are also useful for the first half of the seventeenth century; the survival of some important manuscripts together with the increase in the number of printed books make the period particularly rich in documentary evidence. However, sixteenth-century and seventeenth-century sources differ fundamentally in form and content. Many of the early manuscripts, that is, those dating from the sixteenth-century, contain separate recipes for making pigments or preparing colours, whereas seventeenth-century documents are much more literary in style and take the form of treatises which include comments on the characteristics and uses of various colours but do not generally include recipes for the manufacture of pigments. The alteration in subject matter and style appears to have taken place in the latter part of the sixteenth century after the publication, in 1573, of the treatise on limning which is generally regarded as being the earliest English printed book on the subject of painting. It is true that collections of miscellaneous instructions are to be found amongst manuscripts written after 1600 but, when they are compared with sixteenth-century documents, it is evident that many were copied from earlier works.

Some preliminary discussion of the manuscript sources is essential in order to indicate those which may be considered most reliable and which are, therefore, quoted in the following chapters on pigments. Also, by devoting some attention to the features which are common to several sixteenth-century manuscripts, it is possible to distinguish between those seventeenth-century sources which contain information dating from an earlier period and those which give details of contemporary practice. Whenever possible, technique and authorship are mentioned. The painting technique which forms the subject matter of the various documents must be kept in mind, because pigments which were used in one medium were not always acceptable in another and comments on colours are liable to vary accordingly. Authorship is important, for if an author was a practising painter his comments are likely to be based on first-hand knowledge, although this does not mean to say that famous painters were always the most informative writers. However, the consideration of authorship makes it possible to single out the writers of original material from the many plagiarists active during the seventeenth century.

The early manuscripts are generally anonymous and contain fragmentary information on pigments interspersed amongst other, often related, subjects such as alchemy, medicine and heraldry.[1] The oldest manuscript referred to in the text is B.M. MS. Sloane 122, which dates from about 1500. It includes practical instructions on a number of topics, including gilding and the manufacture of pigments for manuscript illumination. The manuscript belongs to an earlier period than others referred to below, but it is interesting to find that it contains recipes for making whites from lead and tin which are repeated in slightly later sources. The same instructions for ceruse made from tin appear in B.M. MS. Sloane 288, which, although written in a seventeenth-century hand, contains subject matter dating from the first half of the sixteenth century or earlier. Similarly, B.M. MS. Sloane 3292 is written in an early seventeenth-century hand but includes information concerning colours for limning which is dated 1564.

The brief details concerning pigments contained in Bodleian MS. Ashmole 1480 are written in a hand which dates from the sixteenth century, and the subject matter may therefore be assigned to that period or even earlier. Three recipes which it contains, one for making synoper (vermilion) from mercury and sulphur, and two others for allegedly making blue from the same materials, appear frequently in sources of the same period. The same methods for making azure and bice appear in MS. Sloane 288, which has already been mentioned, and a similar recipe for azure appears in Bodleian MS. Ashmole 1494. The manuscript and its companion volume, MS. Ashmole 1491, together form a work on alchemical and medical subjects arranged in alphabetical order. The whole was collated and written by Simon Forman, an astrologer and physician who died in 1611.[2] The repetition of recipes for azure from mercury and white from tin almost ceased after the beginning of the seventeenth century, and their inclusion in a manuscript of that period indicates that parts of it do not describe seventeenth-century practice. A manuscript which comes into this category is B.M. MS. Sloane 1394, which, according to the script, dates from the early seventeenth century, but appears to have been copied from another manuscript of earlier date. It supplies instructions for making various colours, including vermilion, and also the traditional recipe for making ceruse from tin.

A close comparison of colour lists in seventeenth-century sources and slightly earlier works shows that certain changes in nomenclature took place at about the turn of the century. Some names, such as florey, general and ceder [*sic*] green, became obsolete in the early seventeenth century and seldom appear in documents of that period. Azure was no longer used as a colour name but was restricted to heraldic or poetic language, and *indico* became the accepted name for indigo, taking the place of the archaic *Indebaudias* and the intermediate form *Inde Blew*. Thus Simon Forman's colour list in MS. Ashmole 1494, which dates from about 1600–10, has a slightly out-of-date appearance which reinforces the suspicion that the details concerning colours were not original but were copied from an earlier source. Another seventeenth-century manuscript which contains some information dating from an earlier period is B.M. MS. Stowe 680. Some of the details concerning colours are very similar to those in MS. Sloane 3292. However, a recipe for sap green appears to be original, and some other features, such as a distinction between English Inde and Inde Baudias together with the inclusion of sap yellow and fustic yellow, are unusual. It is difficult, therefore, to date the price list, but the inclusion of both verditer and sap green suggests that even if it is not contemporary with the early seventeenth-century hand in which it is written it should not be dated much before the late sixteenth century.

The colour list in B.M. MS. Sloane 6284 appears a little later in date than that in MS. Stowe 680 (*Figure 1*), and it seems probable that it is contemporary with the rest of the book, which is mainly concerned with ceremonial details of royal processions in 1603. The last few pages include a colour list and details of colour mixtures collected by a Mr Rogers the Graver. The inclusion of smalt and the exclusion of general and florey make it likely that the list is a little later than that in MS. Stowe 680, whereas the retention of ceder green and burnt fruit stone blacks suggests that it was compiled quite early in the seventeenth century.

Such are the details which help to determine the age or originality of the information contained in a manuscript. It may be argued that the inclusion of a traditional recipe, such as that for making ceruse from tin, or the inclusion of some traditional colour names does not necessarily mean that a seventeenth-century manuscript was out of date when it was

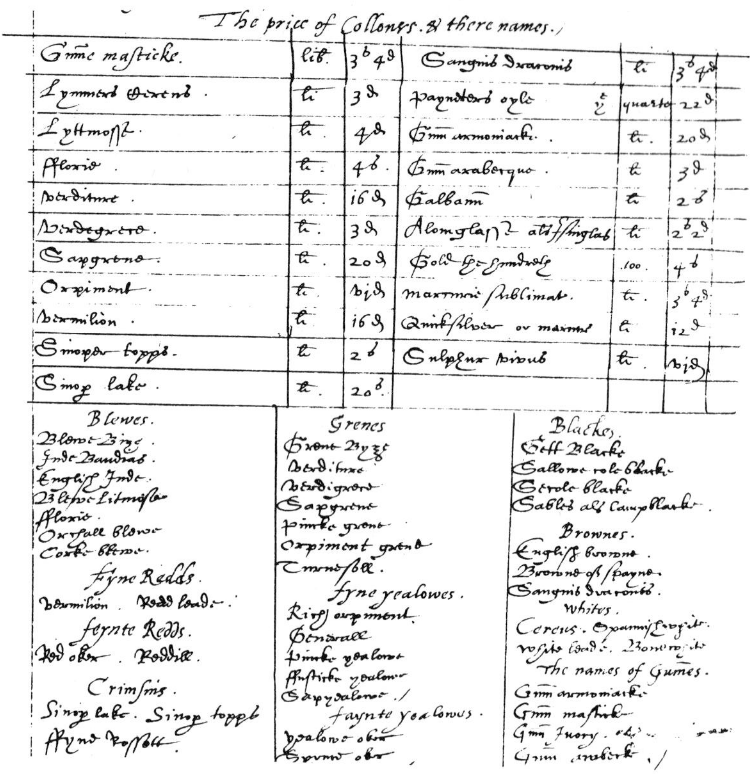

Fig. 1. Early seventeenth-century colour list: sinoper lake is the most costly, listed at twenty shillings per pound, whereas ceruse and verdigris are the cheapest at threepence per pound (British Museum MS. Stowe 680, f.135v). (Reproduced by courtesy of the Trustees of the British Museum.)

written. Nevertheless, comparison with printed books of the same period together with the few manuscripts of known authorship shows that certain colours either were or became obsolete at the beginning of the century.

Before turning to early printed sources and seventeenth-century manuscript treatises, V. & A. MS. 86.EE.69 deserves mention, as, apart from information on water-colour painting and gilding, it includes references to oil painting materials which are attributed to a painter, Edmonde Barton, and dated 1582. Other parts were written later, as they contain details taken from Haydocke's book which was published in 1598. The latter part also includes a colour list (*Figure 29*) with prices which, excepting those for sap green and sinoper lake, are considerably higher than those listed in MS. Stowe 680. There are also details of the price charged for a portrait by Barton. The manuscript holds an important place amongst other sixteenth-century English documents on painting owing to the evidence it supplies concerning English practice in oil painting at that time.

Water-colour painting formed the subject of most early works, including the first book on painting to be printed in English. The book, which was written anonymously and first printed in 1573, has an extremely long title which begins, 'A very proper treatise, wherein is briefly sett forthe the arte of Limming . . .' It is referred to below as *Limming*, 1573.[3] Early editions were well known, for the book's influence can be traced in late sixteenth-century and early seventeenth-century manuscripts through the frequent repetition of a benzoin varnish recipe beginning, 'Take Bengewyn and bray it well betwixt two papers . . .' In the original it is described as 'a kynde of colouring called Vernix wherewith you may vernishe golde, siluer, or any other colour or payntinges, be it vpon velym, paper, tymber, stone, leade, copper, glass &c.' The utility of the all-purpose varnish obviously impressed a large number of readers. Most of the book is concerned, however, with heraldic blazonry and manuscript illumination and it contains certain details, such as instructions for the use of egg yolk as a masking fluid, which belongs to water-colour painting of the sixteenth century and earlier. On the other hand, the book differs from most sixteenth-century manuscript sources in that it is a treatise and not a collection of recipes. It contains instructions for the preparation of colours but not the manufacture of pigments; when the anonymous author states that apothecaries fraudulently mix sand with the pigments, one is led to infer that they were the accepted suppliers.

The second book on painting to be printed in English during the sixteenth century was a translation of the Italian work *Trattato dell'arte de la pittura* by Lomazzo, which was printed as *A Tracte containing the Artes of curious Paintinge, Caruinge & Buildinge* in 1598. The translation was made by Richard Haydocke, a physician, who added some marginal comments of his own. The work is comprehensive, as it covers proportion, colour, light, perspective and the aesthetic side of painting as well as various techniques. Nevertheless, the book is of limited value as a source for the history of pigments in sixteenth-century England, and it is quite clear that Haydocke was uncertain how to translate some of the Italian colour names and may not have known what they were. His influence was far-reaching, however, as many English writers quoted from his book, often without acknowledgement, and Hilliard's treatise was written at his suggestion.

The only known surviving copy of Hilliard's treatise is MS. Laing III 174 at Edinburgh University. At one time it belonged to George Vertue, the eighteenth-century antiquary, who inscribed it 'A Treatise Concerning the Arte of Limning writ by N Hilliard'. The attribution, which is generally accepted, was explained by Vertue and the treatise described as follows:[4]

> The Manuscript of Nicholas Hilliard written with his own hand according to his promise to R. Haydock, who translated & published Lomazzo in English 1598.
> Of Rules & precepts with instructions for limning contained in 28 pages of sheets of paper. The first part concerning the Art of Limning the profession of a Gentleman. The following part also the true description of Jewells of the five different natural Colours.
> This seems to be writ particularly after the publication of Lomazzo whom he often mentions & after the death of Ld Chancellor Hatton whom he mentions & yet before the death of Queen Elizabeth.

The description is generally accurate, although the Edinburgh manuscript is not in Hilliard's hand but is a copy dated 1624, with three

Fig. 2. Nicholas Hilliard, self-portrait at the age of thirty: miniature, signed and dated 1577 (Reproduced by courtesy of Victoria & Albert Museum).

additional pages on limning which are not part of the original treatise. Although a relatively small part of the manuscript is concerned with colours for miniature painting, the complete treatise provides illuminating comment on Hilliard's technique and ideas. An important feature of the work lies in the way it reflects the change in status of painters from artisans to artists which had taken place during the Renaissance period. Hilliard represents water-colour painting as a gentlemanly art; the subject is mentioned at the beginning of the treatise:[5]

> Now therefor I wish it weare so that none should medle with limning but gentlemen alone, for that it is a kind of gentill painting of lesse subiection then any other for one may leave when hee will, his coullers nor his work taketh any harme by it. Moreover it is secreat, a man may usse it and scarsly be perseaved of his owne folke. It is sweet and cleanly to usse, and it is a thing apart from all other Painting or drawing and tendeth not to comon mens usse, either for furnishing of Howsses, or any patternes for tapistries, or Building, or any other works whatsoever, and yet it excelleth all other Painting whatsoever in sondry points.

Famous as this passage is, Hilliard cannot be given all the credit for encouraging a new image for the artist in England. Whereas he maintained that only a gentleman could bring to limning the 'tender sences quiet and apt' which he considered necessary for the art, another writer suggested that the gentleman had much to gain from the practice of painting.

Henry Peacham (1576–1644) was artistically gifted according to his own account, but was a teacher by profession and was at one time tutor to the sons of Thomas Howard, Earl of Arundel, a noted patron of the arts.[6] It is possible that Peacham's writing had greater influence than Hilliard's, because his books on education were printed, some of them in several editions. In the following passage he recommends that the young nobleman should practise washing, that is, topographical painting, which was obviously considered easier than limning (as the art of portrait miniatures was then known):[7]

> I could wish you now and then, to exercise your Pen in Drawing and imitating Cards and Mappes; as also your Pencill in washing and colouring small Tables of Countries and places, which at your leasure you may in one fortnight easily learne to doe; for the practice of the hand doth speedily instruct the minde, and strongly confirme the memory beyond any thing else; nor thinke it any disgrace unto you, since in other Countries it is the practice of Princes, as I have shewed heretofore; also many of our young Nobilitie in England exercise the same with great felicitie.

Thus, as Hilliard had helped to establish the position of the professional painter, Peacham ensured that painting was a respectable pastime for the amateur. Following the publication of his books, others were written specially for beginners and amateurs, so it seems that the number of amateurs must have been quite large during the seventeenth century. At that period, it was still usual for a professional painter to learn about materials and technique by working in a studio, but amateurs were taught by private teachers and books. It is evident that an amateur was not generally expected to prepare his own colours and certainly not the pigments, a circumstance which probably accounts for the disappearance of recipes from most seventeenth-century works on painting.

Peacham's first book, *The Art of Drawing with the Pen,* was printed in 1606 and reprinted the following year. The contents were expanded in *Graphice,* 1612, which was reissued in the same year as *The Gentlemans Exercise. The Compleat Gentleman,* first printed in 1622, is a larger work than the others, but the information on drawing and limning appears in condensed form. It contains additional matter on education and, most important to the reader interested in painting, it includes information on oil painting. The author makes an interesting comparison of painting in oils and water colours, giving reasons why the latter is more suitable for the amateur:[8]

> Painting in Oyle is done I confesse with greater iudgement, and is generally of more esteeme then working in water colours; but then it is more Mechanique and will robbe you of over much time from your excellent studies, it being sometime a fortnight or a moneth ere you can finish an ordinary peece . . . Beside, oyle nor oyle-colours, if they drop upon apparrell, wil not out; when water-colours will with the least washing.

The Compleat Gentleman, which appears to be the first book in English to discuss portrait painting in oils, was reprinted in 1626 and 1627; a second, enlarged edition appeared in 1634 and a third in 1661, but they contain no additional information on painting. Of all the books, the most useful for the history of pigments is *Graphice* or *The Gentlemans Exercise.* Peacham groups pigments together by hue and discusses their use as well as the derivation of colour names. He makes frequent references to ancient writers, such as Aristotle, Pliny and Dioscorides, but his reliance on such authorities confuses him about certain colours, reds in particular, because the meaning of some colour names had altered by the seventeenth century. In spite of that shortcoming, the book contains much more information than Hilliard's treatise, most of it being of an eminently sensible and practical nature, which makes it worthy of the influence and popularity which it once enjoyed.

Excerpts from Peacham are to be found in other seventeenth-century works. B.M. MSS. Harley 1279 and Additional 34120 include a passage concerning 'The practice of that famous Limner Hippolito Donato in drawing a small picture of Christ', which appears in *Graphice.* John Bate, author of *The Mysteryes of Nature and Art,* borrowed so much from Peacham (in some parts the wording is identical) that it is sometimes difficult to distinguish whether a passage has been copied from Bate or Peacham. Unacknowledged extracts from one or other appear in B.M. MSS. Sloane 228 and 1448B and in Bodleian MS. Ashmole 768. Bate's book was printed in 1633 and in the following year with the author's initials (I.B.), but a further edition followed in 1635 under the author's name. Drawing, limning, colouring, painting and engraving form the subject of the third section; other parts are concerned with water-works, pyrotechnics and 'extravagants' such as party tricks and medical remedies. It is doubtful if any of the details concerning painting come from first-hand knowledge, because, apart from the large proportion on colours for limning and oil painting which is derived from Peacham, instructions for making a varnish and a double size come from *Limming*, 1573, a recipe for a purple colour is taken from the *Illuminirbuch* by Valentin Boltz von Rufach, and parts on limning technique are reminiscent of Hilliard.

Returning from printed books to manuscripts of the seventeenth century, one comes to an important group of documents of known authorship. Mention has already been made of two physicians, Simon Forman and Richard Haydocke, who interested themselves in painting, and such interest on the part of doctors is not surprising in view of the close association between medicine and pigments. The Latin word *pigmentum* was used for both pigment and drug, and in the seventeenth century several substances were still used for both purposes. Another physician who interested himself in pigments and painting was Sir Theodore Turquet de Mayerne (1573–1655), a native of Geneva who lived in England from 1610 and practised at the court of the early Stuarts. His position there gave him the opportunity to meet many eminent painters, including Rubens and Van Dyck, and to collect information from them. In recording details of painting materials and technique of the time, de Mayerne made an important contribution to the history of art, but this must not be allowed to overshadow his greater contribution to the history of medicine. At a time when some medical practitioners were noted for their extra-medical activities (Forman achieved notoriety for witchcraft and Haydocke drew attention to himself through an elaborate hoax of preaching in his sleep), de Mayerne was undoubtedly competent. He is now famous for his businesslike, almost modern approach to the day-to-day recording of case histories.[9] His practical outlook and attention to detail are evident in many of his manuscripts, including B.M. MSS. Sloane 1990 and 2052, which contain notes on the subject of painting.

Fig. 3. Sir Theodore Turquet de Mayerne, 1573–1655: portrait miniature in enamel by Petitot (Reproduced by courtesy of National Portrait Gallery).

In past years, MS. Sloane 2052 has been the better known of the two manuscripts as it has appeared in print.[10] Berger's edition contains a transcript of the complete manuscript in the order in which it was originally written, accompanied by a translation and notes in German. Van de Graaf's edition includes a transcript of those parts of the manuscript which are written in de Mayerne's hand but excludes the various notes and letters written by others. An important feature of the edition, which contains an introduction and notes in Dutch, is the rearranged order; all the information on supports, priming, pigments, oils and varnishes is grouped together according to subject matter. Thus, in place of the confused nature of the original, which was written as a private notebook over a period of about twenty years beginning in 1620, one has a collection of notes and recipes arranged in a logical order which makes the comparison of entries on a particular subject very much easier. A large part of the manuscript is concerned with the materials and technique of oil painting, but there are some details concerning painting in distemper and water colours. The book also contains pages of named colour samples (see *Plates* 7 and *8*). Much of the evidence is extremely valuable, because de Mayerne frequently inserted a note concerning the source of his information and it is clear that much was obtained from professional painters working at the English court. There is, nevertheless, a certain amount of evidence of doubtful value, such as the group of recipes for making blue from mercury, which suggests that some information was copied from old manuscripts. De Mayerne had, however, experimented with the manufacture of pigments, and the manuscript includes comments on the success or failure of his attempts to follow various recipes.

The other relevant notebook compiled by de Mayerne, MS. Sloane 1990, has in the past been better known for an important passage on soap-making than for its contents on painting.[11] Again, most of the manuscript is in de Mayerne's hand, and it contains notes compiled over a similar period, having entries dated between 1623 and 1644. The contents of the last forty folios are not by de Mayerne but were written at a latter date by another, possibly John Colladon, also a native of Geneva resident in England, to whom de Mayerne bequeathed his manuscripts. The book covers a greater number of subjects than MS. Sloane 2052, including details on the art of painting, etching, sculpture, glass, enamelling, ceramics, metalwork and dyeing, and miscellaneous details such as recipes for medicine and confectionery. The part on miniature painting in enamels is important for the history of pigments, because it suggests that certain colours were used in that art before they were adopted in oil or water-colour painting. It is reasonable to suppose that much of the information on the subject was gathered either by or for Jean Petitot, a compatriot and friend of de Mayerne, who may be considered as one of the greatest exponents of portrait miniature in enamel. He is mentioned by name in the manuscript and is known to have painted a miniature portrait of de Mayerne.

The group of manuscripts of known authorship comprises both of de Mayerne's notebooks and is completed by the various copies of the treatise on miniature painting in water colours which was written by Edward Norgate (1581–1650) at the request of de Mayerne. No autograph copy appears to have survived, and only one, posthumous, copy bears Norgate's name. Nevertheless, the acquaintanceship between Norgate and de Mayerne is established, for Norgate's name is mentioned as a contributor to the notes in MS. Sloane 2052, and de Mayerne's request for a treatise on limning is recorded at the beginning of the later version of Norgate's treatise, in which the author states patriotically:[12]

> . . . at the request of that learned Phisitian, Sir Theodor Mayerne, I

> wrote this ensewing discourse. His desire was to know the names, nature and property of the severall Colleurs of Limming, comonly used by those excellent Artists of our Nation who infinitely transcend those of his.

The authorship of Norgate has been accepted since the publication of a transcript of the later version in 1919, edited and with an introduction by Martin Hardie, who points out parallels between contemporary references in the treatise (mention of paintings seen during a trip to Italy, for example) and Norgate's career. Even so, there are so many extant copies of the treatise, some of which are or have been attributed to other authors, that it is worth listing the manuscripts in detail.

According to the internal evidence which is discussed by Hardie, the first version of Norgate's treatise appears to have been written *circa* 1621–26.[13] B.M. MS. Harley 6000 dates from about that period and is the most accurate and best written of all the copies of the early version. The full title of the manuscript is 'An exact and Compendious Discours concerning the Art of Miniatura or Limning, the names, Nature and properties of the Coullours, the orders to be observed in preparing and using them both for Picture by the Life, Landscape and Historyes'. B.M. MS. Additional 23080 is also a copy of the early version; it is written in a late seventeenth-century hand with additions in a later hand, almost certainly that of Vertue, which are taken from William Sanderson's *Graphice,* published in 1658. Vertue owned the manuscript but was mistaken concerning its authorship, as he wrote in one of his notebooks:[14]

> . . . in a MS. Treatise on Limming writ by . . . a dear Cousin of Isaac Oliver . . . Sr. Wm. Sanderson who printed it partly—1658 fol. An old Coppy of this MS. I have had many years by me which I had descended by—Russel from Olivers relations.

He was confused, as others have been, by the author's allusion to Isaac Oliver as his cousin. Sanderson, who printed part of the treatise, as Vertue says, was not his cousin any more than Norgate was. As Hardie points out, 'cousin' was a seventeenth-century term of endearment which was not necessarily connected with kinship.

It is interesting that Vertue's copy, MS. Additional 23080, came to him from the Russell family because Theodore and Anthony Russell were both painters and related to Isaac Oliver by marriage. On account of the relationship, that part of Bodleian MS. Rawlinson D. 1361 (ff. 205–39) which contains a copy of Norgate's earlier treatise written in a mid-seventeenth-century hand is tentatively attributed to the Russells in the printed catalogue of the Rawlinson manuscripts. Neither that manuscript nor Bodleian MS. Ashmole 768 is mentioned by Hardie in his discussion of copies of Norgate's treatise, and it is probably for that reason that the generally accepted attribution to Norgate has been overlooked in both cases. In the printed catalogue of the Ashmolean manuscripts, it is suggested that that part of MS. Ashmole 768 which contains information about painting (pp. 408–84) may have been written by Elias Ashmole when young. It is possible that he made the copy, but most of it is a transcript of the first version of Norgate's treatise with additional colours obviously copied from Haydocke. Certain pages are copied from Bate without acknowledgement, and others contain notes on colours which are acknowledged as being copied from Lomazzo, presumably Haydocke's translation.

Other manuscripts which contain copies of the early version are B.M. MSS. Sloane 228, Additional 12461 and 34120, all of which were written in the seventeenth century, and V. & A. MS. 86.FF.19, which is an eighteenth-century copy. All of them contain some additional information. For example, MS. Additional 34120 contains recipes from an earlier period, and MS. Additional 12461 contains miscellaneous information about oil painting and varnishes and includes details of the flesh colours used by Van Dyck and Lely. Much of the same information is included in the Victoria & Albert Museum manuscript, including the same brief description of the methods of the painters mentioned above. Such manuscripts are interesting on account of the additional matter included, but some of them contain bad errors of transcription in the main part on miniature painting and they are not particularly valuable as sources for Norgate's treatise.

Undoubtedly the most important copies of the treatise are MS. Harley 6000, Bodleian MS. Tanner 326 and Royal Society MS. 136. As already explained, the first mentioned is the earliest and dates from the 1620s; the others are copies of the revised treatise which was probably composed between the years 1648 and 1650. The first version differs from the second in that it contains more details about materials whereas the later version has greater emphasis on the last part and contains more comments on the aesthetic side of painting. The copies of the later version differ very little, although MS. Tanner 326 has some additions in a different hand which are not original but were copied from Bate's *Mysteryes.* The attribution to Norgate is supported by both copies of the later version. MS. Tanner 326, which includes the intials E.N. at the end of the dedication, was probably copied out soon after Norgate's death in 1650. It has always been attributed to Norgate since its acquisition by the Bodleian Library in 1735. The Royal Society manuscript, which is dated 1657, is the only one in which the words 'By E. Norgate' appear after the title. Its authenticity can hardly be questioned because the second version of the treatise was dedicated to Henry Howard, third Earl of Arundel, and the manuscript was in the Arundel library until 1678 when it was transferred with many other volumes to the Royal Society. Norgate was well known to the Howard family, as he owed his official appointments at the College of Arms to Thomas Howard, second Earl of Arundel, acting in his capacity as Earl Marshal, and also private appointments as purchaser of paintings for the earl and as teacher of painting to his sons. It is most improbable, therefore, that the Royal Society Arundel manuscript would have been attributed to Norgate if he was not actually the author.

Norgate was well fitted to write the treatise, because his work as a

Fig. 4. Judith Norgate: portrait miniature of his wife by Edward Norgate, dated 1617 (Reproduced by courtesy of Victoria & Albert Museum).

Herald and later as Clerk of the Signet required much writing as well as some painting and he is also known as a painter of portrait miniatures. The very great value of the work lies in the fact that it is instructional and that no previous knowledge is assumed. It includes full details of the way in which the pigments and the painting support must be prepared, and it gives stage-by-stage instructions in portrait painting. Those parts are followed by sections on landscape and history painting in miniature, and at the end are instructions for making ultramarine, making and using crayons, and, in the later version, instructions for making pink. Thus the treatise contains details concerning the manufacture of very few pigments, but the comments on the characteristics of various colours are detailed and the sections on technique are so well explained that the whole work is exceptionally useful. Apart from the author's first-hand knowledge of the subject, the treatise contains references to the work of Hilliard, both Olivers, and Hoskins, and it may be regarded as an accurate account of English practice in miniature painting during the first part of the seventeenth century. It is now extremely valuable as an historical source, but it was also highly valued during the seventeenth century as it stands out as the most copied English work on painting of that time. In addition to the manuscripts mentioned above which contain all or most of the treatise, some printed books, which will be mentioned in the next chapter, also contain large portions of Norgate's work.

Another manuscript which contains much of Norgate's treatise is B.M. MS. Harley 6376, which betrays the fact in the title, 'The Art of Limning either by the Life, Landscape or Histories'. It differs from other copies, however, in that a discussion of drawing materials and painting tools is included before the part on colours and technique, which follows Norgate fairly closely but with numerous additional comments. A section entitled 'The Art of painting in Oyle by the Life' is dated 1664. The remainder is written in a later hand and includes notes on water and oil gilding and the composition of varnishes which include the dates 1679 and 1683 and were obviously written at various times. The name of Henry Gyles is written at both the beginning and end of the book; it is clear that it belonged to him and it is possible, although not certain, that he wrote the entire manuscript. The question of authorship is of some importance, because the part on oil painting materials and technique is original and other parts concerning tools which are inserted at the beginning bear no relation to Norgate's treatise. The sketches of equipment are also original; they are not copied from Bate's *Mysteryes,* which is the only earlier book with similar illustrations. It seems likely that, if the original parts had concerned glass-painting instead of oils, they would have been attributed to the owner, Henry Gyles (1645–1709), because he is famous as a glass-painter of York. Little is known of Gyles' early life, which is an important period as far as authorship is concerned, but there is considerable information about his later years. Comparison of references in the last part of MS. Harley 6376 with other sources makes it virtually certain that Gyles wrote that part. Vertue knew of the manuscript and wondered about its authorship, singling out the point which could settle the matter:[15]

> . . . book. MS. of the Art of Limning—which did belong to Henry Giles Glass painter—of York.—
> who it was writ by (I cant find) but suppose before 1660 the writer in p. 5. thus
> for I remember when I did learne to draw before I did draw well, I desired to learne to paint; but my Master Mr. Wm. Martins ye Elder answer'd me very wisely, that I must not run before I coud go.

Unfortunately, insufficient is known of Gyles' youth to tell if he was taught by someone called Martins and to say whether or not he had a comprehensive training in all branches of painting, including oils. Nevertheless, much of the information contained in MS. Harley 6376 is original, and in the following chapters its authorship is attributed to Gyles.

The contents of Gyles' book were transcribed at an early date, and a manuscript copy exists in the Victoria & Albert Museum, MS. R.C.A. 20. It is carefully written and decorated in red ink, but it contains none of the later notes written after 1664 and it lacks the pencil sketches which form such an interesting part of the original.

Another seventeenth-century book, B.M. MS. Egerton 1636, also contains thumbnail sketches of painting equipment. It is a pocket-size

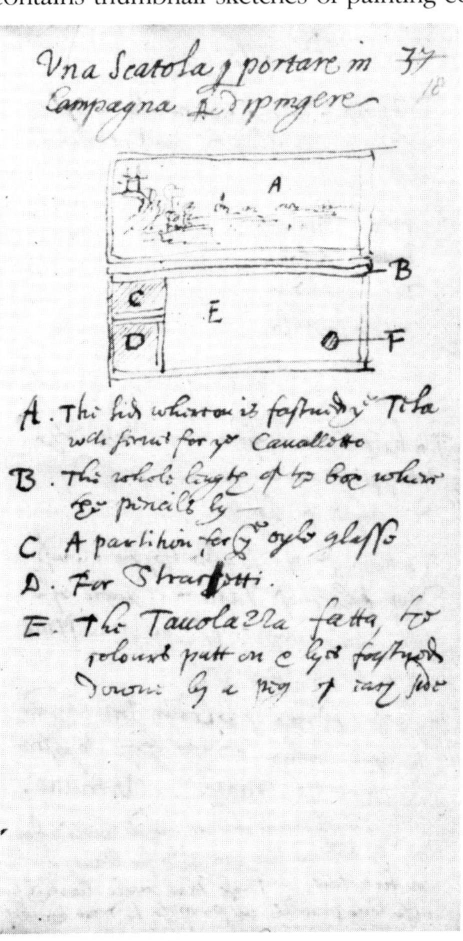

Fig. 5. Diagram of an oil colour box for sketching out of doors by Richard Symonds when in Italy, 1650–51 (British Museum MS. Egerton 1636, f.18). (Reproduced by courtesy of the Trustees of the British Museum.)

notebook compiled by Richard Symonds (1617–92?) mainly during the years 1650–51 when he was in Rome.[16] It is written in Italian and English, and would seem to be more important as a source for the history of painting in Italy than in England because most of the information is attributed to Giovanni Angelo Canini. Amongst other details, the book contains comments on priming, setting a palette and working with various colours, and a section is devoted to a stage-by-stage account of the painting of a large picture of Antony and Cleopatra by Canini. That painter was active in the Academy of San Luca at Rome, and the notebook contains some diagrams of the type of equipment used there. Also, it is interesting to find the layout of an outdoor oil sketching box with a sketch of an Italian scene held in the lid, thus suggesting that Italian *veduta* painting may have been well established somewhat earlier than is generally supposed. From the purely English point of view, the book provides interesting evidence of Italian influence in England in the seventeenth century, for, whereas Norgate visited Italy to arrange for the purchase of works of art, Symonds returned from Italy with information about Italian methods. It is tempting to suppose that Symonds went to Rome specifically to obtain an academic training which was then unavailable in England, but such is not the case as he was not a professional painter but a militant royalist who found absence from England desirable during the early years of the Commonwealth. Nevertheless, he appears to have spent much of his time gathering information about art and art training in Italy. The full impact of Italian art on English painting was not felt until the following century, but it is interesting to find seventeenth-century manuscript sources which provide evidence of active interest in Italian art on the part of painters and connoisseurs alike.

2

Literary Sources: Printed Books 1650–1835

Painting, staining, limning and washing indicated four different types of painting during the seventeenth century. In modern usage *painting* means painting in any medium, but in the past it was applied to oil painting as opposed to water-colour painting and it is frequently used in that sense in seventeenth-century books. At that time it included any easel painting in oils, no matter what support was used, although at an even earlier period a painter was one who painted on wood, leather or plaster—in fact on any support but textiles. Fabric painting was referred to as staining, which was in its turn quite distinct from dyeing. During the medieval period the competition between craftsmen painters and stainers was intense, each company guarding its right to paint on a particular type of support. By the seventeenth century the situation was not quite the same, the two groups being combined as one, the Painter–Stainers Company, but the use of staining to mean decorative painting on textiles and painting to indicate oil painting, either decorative or artistic, was still current. Thus, William Sanderson's *Graphice,* 1658, is divided into two parts: the first is entitled 'The Use of the Pen and Pencil in the most excellent Art of Painting', implying that it is concerned mainly with oil painting, whereas the second deals with the 'art of limning in water colours'. During the seventeenth century limning meant miniature painting and did not include the other method of water-colour painting, known as washing, whereby broad washes of colour were applied (as, for example, in the colouring of prints and maps). Sanderson included some original remarks on contemporary oil painters in the first part of his book, but he copied the second part from Norgate with the result that it includes nothing about washing, a subject which is mentioned quite frequently in other books dating from the second half of the seventeenth century.

Another writer who borrowed from Norgate was Alexander Browne. On the title page of *The whole Art of Drawing, Painting, Limning and Etching,* 1660, and the enlarged version, *Ars Pictoria,* 1669, he describes himself as a practitioner, but it is probably as a writer and a teacher of water-colour painting, especially to Mrs Samuel Pepys, that he is best known.[17] The first part of *Ars Pictoria* is almost entirely Haydocke's

Fig. 6. Alexander Browne: engraving by de Jode after portrait by Huysmans, from the frontispiece to *Ars Pictoria*, 1669. (Reproduced by courtesy of the Trustees of the British Museum.)

translation of Lomazzo rearranged in a different order. The author acknowledges a debt to Hilliard, but the second part owes much to Norgate. Even though a large part of the contents is copied from other writers, the book is not without interest for the history of pigments, because by the time it was reprinted in 1675 Browne had changed his mind about the usefulness of certain colours and it contains some informative alterations. Additionally, there is evidence that Browne extended his activities and set up as a colourman, a business which he doubtless built up through his private pupils. Some details are given in an advertisement which Browne inserted in the second edition of *Ars Pictoria*:[18]

> Because it is very difficult to procure the Colours for Limning rightly prepared, of the best and briskest Colours, I have made it part of my business any time these 16 Years, to collect as many of them as were exceeding good, not onely here, but beyond the Seas. And for those Colours that I could not meet with all to my mind, I have taken the care and pains to make them my self. Out of which Collection I have prepared a sufficient Quantity, not onely for my own use, but being resolved not to be Niggardly of the same, am willing to supply any Ingenious Persons that have occasion for the same at a reasonable rate, and all other Materials useful for Limning, which are to be had at my Lodging in Long-acre, at the Sign of the Pestel and Mortar, an Apothecary's Shop; and at Mr. Tooker's Shop, at the Sign of the Globe, over against Ivie Bridge in the Strand.

There are earlier references to the retail sale of artists' colours in seventeenth-century England, but this is possibly the earliest extant advertisement on the subject.

Several other books of the period were addressed to amateurs and young people, many of whom practised water-colour painting although not always with good results. One author lists common faults after remarking: 'For as much as Colouring Prints, and Maps, is of common use, and much Practised by the Gentry and Youths, who for want of knowledge therein, instead of making them better, quite spoyl them.' The passage occurs in *Academia Italica, the Publick School of Drawing, or the Gentlemans Accomplishment,* which was published in 1666 with the author's initials, T.P. The book is divided into two parts, the first on water-colour painting, the second on oil painting. Bound with the copy belonging to the Victoria & Albert Museum is another book of the same date and similar size and appearance but with the separate title *A Book of Drawing, Limning, Washing or Colouring of Maps and Prints.* Its authorship is anonymous, perhaps advisedly, as the first two pages are lifted entirely from *Limming,* 1573, a circumstance which makes one suspect that other parts may not be original. Another anonymous work is *The Excellency of the Pen and Pencil,* printed in 1668 and again in 1688. The title page includes an acknowledgement that the information was collected from the writings of 'the ablest Masters both Antient and Modern'. The section on miniature painting seems reminiscent of Hilliard and Norgate, but as everyone seems to have recommended the use of abortive parchment and the necessity for three sittings with the sitter placed in a north light, it becomes difficult to distinguish plagiarism from tradition. Following that section is another on oil painting, including portrait painting from life and picture cleaning, and a final section contains instructions for washing prints and maps. That book certainly contains some original remarks, but one which may be dismissed as a complete work of plagiarism is *Polygraphice* which first appeared in

1672. The first edition gives only the author's initials, but the many later editions give the author's name as William Salmon. The preface contains an admission that the author had made use of other works. Much of the first part on drawing was taken from Peacham, instructions for limning come from Norgate, and chapters concerning 'experimental observations' are an adaptation of parts of Robert Boyle's *Experiments and Considerations touching Colours*. It seems likely that the final chapters on gilding and dyeing were also taken from other sources.

It is a change from constant repetitions of Norgate's treatise to encounter a work of quite different origin. A book entitled *An Introduction to the General Art of Drawing . . . likewise an excellent Treatise of the Art of Limning* was published in 1674 with the explanation that it was translated from a Dutch book by W. Gore, who had in turn adapted a book by the miniature painter Mr Gerhard of Brugge. Parts of the work date from the first half of the seventeenth century, for Geerard ter Brugge's treatise on miniature painting, *Verligtery Kunst-Boeck,* was published in 1634. It was later adapted and augmented by Willem Goeree in *Verligterie-Kunde, of regt Gebruik der Water Verwen.* The book was of international importance, as it was translated into both German and English. The anglicised name, Gore for Goeree, was used in the English edition. The Dutch original contains a comprehensive list of colours for limning and a discussion of all of them, but some of those comments are omitted from the English translation. Nevertheless, the English version is quoted in the following chapters on pigments because some of the comments summarise general opinion which was often less well expressed by other writers.

Fig. 7. Willem Goeree: portrait miniature by Peter Cross, signed and dated 1670 (Reproduced by courtesy of Victoria & Albert Museum).

Oil painting is discussed in *The Art of Painting* by John Smith, published in 1676. The book is probably better known as *The Art of Painting in Oyl,* which is the title of the expanded second impression of 1687 and subsequent editions. The third edition of 1701 contains additional matter on the art of colouring maps in water colours. In regard to the part on oil painting, it should be noted that John Smith was a clock-maker, and, consequently, a large part of his book is concerned with the painting of sun-dials and other outdoor objects. This point has to be kept in mind when considering his discussion of colours, as his procedure is likely to have differed from that of a portrait painter, for example. An interesting feature of the book is that, not only does the author write from his own experience, but in his references to the *Philosophical Transactions* of the Royal Society and Lemery's *System of Chymistry* he reflects an awareness of scientific writings of the seventeenth century. Most of the recipes and manufacturing instructions are quoted from such sources, a fact which is acknowledged in the text. It is a disappointing feature of many of the books which went through several editions that, although they are said to be greatly augmented and altered, the later editions seldom include any of the new colours which one could reasonably expect to find. The later editions of John Smith's book do not contain any reference to Prussian blue, although it is included in *A short and direct Method of Painting in Water-Colours* which was printed in 1730 under the authorship of 'the late Mr. Smith'. The book is sometimes listed amongst those by John Smith, but it differs from them in both content and style and it seems most improbable that it was written by the same author.

Another seventeenth-century book on oil painting was written by an author with the same surname. *The Art of Painting according to the Theory and Practise of the best Italian, French and Germane Masters* was published in 1692 under the initials M.S., but the name Marshall Smith appears at the end of the dedication. As the title indicates, the book is concerned with fine art and much of the information was compiled from foreign sources; the author mentions Lomazzo, Vincent and Testling as authorities whom he used. The first few chapters are devoted to a discussion of the status of painting in antiquity and the seventeenth century. It seems likely that Chapter V, entitled 'That this Art is Requisite to the Education of a Gentleman', owes much to Peacham. Several chapters deal with drawing, proportion, light and perspective, and not until half-way through the book is the subject of colours mentioned. The characteristics of a number of pigments are discussed, and advice is included on the most suitable medium for each one. The author differentiates between the preparation of colours for various purposes. For example, fine lake may be ground in linseed oil for ordinary purposes, but nut oil should be used when the colour is needed for a face. The author offers no explanations as to the characteristics of different oil media, and the unfortunate beginner is left to himself to discover that nut oil has less tendency towards yellowing than linseed oil and is therefore more suitable for flesh colour. It is possible that the deficiency arises from the fact that the part on technique is quite short, but the author finds room for some original comments on studio heating and the desirability of weighting easel legs with lead to keep the easel steady—remarks which do not appear in other English works of the same period. The practical nature of the remarks suggest that the author practised painting, possibly as an amateur, for he describes himself as Marshall Smith, Gentleman. The same name and description is included in letters patent, dated 1718, for the invention of a grinding mill which was described as follows:[19]

> A new machine or engine for the grinding of colours, to be used in all kinds of paintings, also for grinding of looking glass, chocolate, and polishing of marble, &c., pounding and sifting of colours, oar, and all hard substances, or other things usually ground with a muller on a flatt stone, by mocion in the mechanicks, exactly imitating those of the handes.

There is no further information about the inventor nor a description of the mechanics of the mill, because Smith's invention was patented before it became customary to enter a specification. It seems probable, however, that the inventor and the author of *The Art of Painting* were one and the same.

References to Italian authorities appear with growing frequency in English books of the late seventeenth century. Both Alexander Browne and Marshall Smith drew on Fialetti for parts of their books which concern drawing, and, as mentioned previously, both took much of their information from Lomazzo. The same authority and other Italian writers were used by John Elsum in his book *The Art of Painting after the Italian Manner,* which was published in 1704. The reference to Naples yellow in his colour list is possibly a reflection of Italian influence.

Few books on painting appear to have been published in England during the first half of the eighteenth century, and they are not distinguished for originality. The anonymous *The Art of Painting in Miniature,* first published in 1729, came entirely from a French work. The author claimed to have a French manuscript on miniature painting in his possession, but it would have been possible for him to have translated one of the printed copies of the *Traité de mignature* which contains the same subject matter. The French book was printed several times between 1674 and 1711. There is a certain amount of confusion concerning the author, owing to the fact that the treatise contains a dedication ending with the initials C.B. and many of the editions were printed by Chrisophe Ballard whose initials are identical. For that reason, it is sometimes attributed to Ballard; for example, Mrs Merrifield refers to it by that name in *Original Treatises . . . on the Arts of Painting.* However, it is here referred to by the name of Boutet, to whom it is attributed by the major libraries.[20] The first English edition of 1729 does not contain some of the recipes which appear at the end of the second French edition of 1674 whereas they are included in the second English edition of 1730. In general, very little was altered by the English translator, and, although the book is useful for a comparison between English and French practice during the second half of the seventeenth century, the English translation cannot be used as a reliable source for the eighteenth century.

The Art of Drawing and Painting in Water-Colours was published under anonymous authorship in 1731 and kept on reappearing either under its own title or in borrowed form under other titles throughout the eighteenth century and even later. It is referred to below by the short title *The Art of Drawing,* 1731. The name of Robert Boyle, the seventeenth-century scientist and Fellow of the Royal Society is sometimes linked with the work, as the anonymous author, who described himself as a teacher of painting, claimed that Lord Carleton had given him an unpublished manuscript written by Boyle. It is quite possible that Carleton, who was related to Boyle and died without heirs in 1725, left a manuscript to the author, but the statement that the papers were unpublished is inaccurate as they obviously contained the same reports of experiments which appear in Boyle's *Experiments and Considerations touching Colours.* This book, which was first printed in 1663 although the earliest extant copies are dated 1664, is important because it contains a discussion of the nature of colour and light which precedes Newton's work on the same subject. However, it was certainly not written for the benefit of painters and it is of minor importance for the history of pigments, its main feature in that respect being a discussion of organic dyes. Unfortunately, Boyle thought that some, such as a blue dye from cornflowers, were light-fast, and the author of *The Art of Drawing* perpetuated the idea. His book is concerned mainly with the colouring of prints, and, for that reason, transparency is accepted as a criterion of a good colour; consequently, many organic dyes are recommended. Owing to numerous reissues, the book's influence continued throughout the eighteenth century. The third edition of 1732 contains additional information, including a chapter on the nature and use of dry colours which, together with instructions for making alum water and lime water, are taken from *A Book of Drawing,* 1666. Further editions appeared in later years with a number of alterations which brought it up to date in detail but not in ideas. According to his own account, the same author wrote *The Art of Drawing in Perspective* and produced *The Art of Painting in Miniature,* which has already been mentioned. All three works were published together in one volume under the title *Arts Companion or a new Assistant for the Ingenious.* Some of the subject matter reappears in the *Dictionarium Polygraphicum,* 1735, which was compiled by John Barrow. The dictionary contains miscellaneous information concerning most of the visual arts. A price list of colours is included in the section on mezzotint.

Any impression of complacency concerning the permanence of most organic colours left by the author of *The Art of Drawing,* 1731, is dispelled by the forthright condemnation of cornflower blue and other dyes which was made by John Hoofnail in *New Practical Improvements . . . on some of the Experiments . . . of Robert Boyle* in 1738. Apparently, Hoofnail had come into possession of Boyle's manuscript but he was disappointed to find that it included only experiments and no detailed recipes. His book is composed mainly of quotations from Boyle with a commentary on the experiments which he had carried out himself. The book also contains an English translation of the instructions for making Prussian blue which were published in Latin in *Philosophical Transactions* in 1724. It appears that John Hoofnail was a man of good education with some scientific training, and it is disappointing that nothing more is known of him.[21]

Returning to oil painting, a book which was influential during the second half of the eighteenth century was *The Practice of Painting* by Thomas Bardwell (1704–67). The author was an English portrait painter and his book, which was devoted to oil painting and perspective, was first printed in 1756. The discussion of colours is limited to about fifteen which were considered most useful in oil painting, and, consequently, the information concerning pigments is somewhat restricted. However, it holds a place of some importance amongst books on painting technique as it has been possible to compare the painter's described and actual technique, and parts of the book were repeated many times by other writers throughout the century.[22]

Much more comprehensive information on pigments is contained in *The Handmaid to the Arts,* the first edition of which was published in 1758 and the second in 1764. Both editions bore only the author's initials, R.D., those of Robert Dossie (1717–77) who also wrote other books on chemistry and was a supporter of the Society of Arts. His interest in

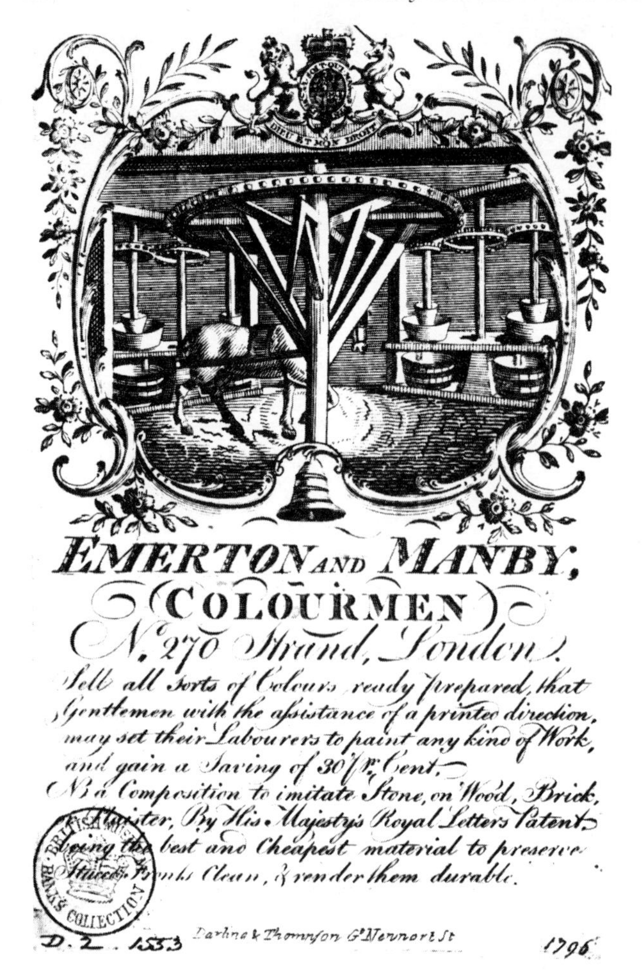

DIEU ET MON DROIT
EMERTON AND MANBY,
(COLOURMEN)
No. 270 Strand, London.
Sell all Sorts of Colours ready prepared, that Gentlemen with the assistance of a printed direction, may set their Labourers to paint any kind of Work, and gain a Saving of 30 Pr. Cent.
NB a Composition to imitate Stone, on Wood, Brick, Plaister, By His Majesty's Royal Letters Patent, being the best and Cheapest material to preserve Stucco Fronts Clean, & render them durable.
BRITISH MUSEUM BANKS COLLECTION
D.2.1553
Darling & Thompson Gt. Newport St
1796

Fig. 8. An eighteenth-century colourman's trade card. Emerton prepared artists' and decorators' colours; those required in large quantities for house-painting were ground on horse-powered mills (British Museum Banks collection, D.2.1553). (Reproduced by courtesy of the Trustees of the British Museum.)

manufacture is reflected in his discussion of the different equipment, such as horse-powered grinding mills illustrated in *Figure 8*, and processes employed in making colours. He deplores the way in which earlier writers copied from others without any first-hand knowledge of the subject, and the fact that pigments were so poorly prepared. Sixty-five colours are included in Dossie's list, and he discusses the preparation and properties of most of them at some length. They are grouped together by hue and are to be found in the first half of the work, the second part containing information on invisible inks, engraving and porcelain, and an appendix with an English translation of the Latin instructions for making Prussian blue. The book is not without minor errors, some of which may have crept in because the author was not a practising painter, although they in no way diminish its importance. Dossie's understanding of the subject of colour-making was possibly equalled by Hoofnail's knowledge, but the scope of his book was very much wider than that of Hoofnail's comparatively small volume. Consequently, Dossie's work stands out as the most authoritative on the subject of artist's colours of the mid-eighteenth century.

During the second half of the eighteenth century the number of books on painting multiplied, but their usefulness did not increase in the same way. Many of them do not have the author's name; they were almost certainly composed of information obtained from earlier sources, and they were often printed for or published and sold by booksellers or printsellers who were sometimes also retail colourmen. Such was the case with *Bowles's Art of Painting in Water-Colours.* Much of the information which it contains comes from one of the later editions of *The Art of Drawing*; other details were probably taken from elsewhere, because the book gives an impression of utter confusion. Some colours suggested in colour mixtures do not appear in the colour lists and there are no details of their composition. The name of Bowles, who published and sold the book, was omitted from the title of later editions, the eighth of which appeared in 1786 and the eighteenth in 1818. Similarly, Carington Bowles published a book called *The Artist's Assistant,* which originally had his name in the title although the fifth edition of 1788 does not include it. That book suffers from the same confusion as the *Art of Painting in Water-Colours.* The somewhat indiscriminate list of colours given in those two books is not repeated in *Bowles's Florist,* but instead fourteen principal water colours are listed, all of them obtainable from the author's shop. The main feature of the book is the collection of sixty plates of flowers for colouring.

More useful is *The Artist's Repository and Drawing Magazine,* which was originally issued in four books between 1784 and 1786. Volume II, *A Compendium of Colors,* is sometimes catalogued as a separate work. The *Compendium* contains some interesting information arranged in alphabetical order of subjects, which are not restricted to colours but include all painting materials. The anonymous author appears to have been well informed, and the comprehensive nature of the information is enhanced by the inclusion of prices current in the 1780s. *The Young Artist's Complete Magazine* Volume I, *circa* 1785, also includes a comprehensive list of colours with their prices. It contains information on drawing and colours, much of which comes from *The Handmaid to the Arts,* referred to by the author in his discussion of the first colour. He also refers to the French encyclopaedia edited by Diderot. Another anonymous work is *The Artist's Assistant, in . . . the Mechanical Sciences,* which is undated but may be about 1790. It contains information on numerous subjects, from perspective to papier mâché, but most of the details about colours are taken from *The Handmaid to the Arts* without any acknowledgement. Details of colours included in *The Art of Painting in Miniature,* 1798, by John Payne are limited because of the restricted subject matter, but at least all the information is original and therefore quite useful.

As previously mentioned, a number of eighteenth-century writers on oil painting borrowed from Bardwell, but one which appears to be original is *An Essay on the Mechanic of Oil Colours* by W. Williams, printed in 1787. The book is not very long, and the remarks concerning colours are mainly restricted to points which an amateur should know when using ready-made colours. A larger part of the book is concerned with oil painting media, and it concludes with a number of recipes which are strangely incomplete in that the names of the constituents have been omitted deliberately and blank spaces left, possibly for the key words to be inserted later. For example, the words 'twelve ounces of pure *' have the following note appended:[23]

> *And here again look after the druggist, or you will have half or as it is much cheaper than but will not dissolve in without more heat than will bear, nor answer the end when done.

The recipes purport to be instructions for the preparation of colourless driers and an oil-based picture varnish. They were offered to the reader as 'valuable secrets', because the author, who was resident in Bath, maintained that it was impossible to purchase a good, colourless drying oil from the colourmen. He recommends that colours should be ground in poppy oil because of its nearly colourless nature and warns that a newly painted picture should not be varnished with anything which dries more quickly than the colours, but other advice is based on less sound ideas. His instructions for heating oils and other constituents would be dangerous for an amateur to carry out owing to the inflammable nature of the materials, and his remark concerning an oil varnish, 'it is not to be taken off again, as many weak connoisseurs think all varnishes should, but it is a mistaken notion', is not in accordance with modern ideas.

A much more extensive work is the *Practical Treatise on Painting in Oil-Colours,* which was published in 1795. The anonymous author includes part of Bardwell's *The Practice of Painting,* to which he refers in his introduction, but he adds a comprehensive discussion of *materia pictoria,* which includes colours. The following part, which concerns the technique of oil painting, also contains useful comments on colours, as do the miscellaneous observations further on. The intervening part includes remarks on varnish-making and a summary of *Coloritto,* a work on colour mixing by Le Blon, which in its English form was probably taken from Bardwell. At the end is a section containing an English translation of the paper on white pigments written by the French chemist Guyton de Morveau. The *Practical Treatise* is a useful book which probably had some influence, as much of the additional part on oil painting in later editions of *A Compendium of Colors* is identical with information given in the treatise.

Some of the books mentioned so far refer to colourmen or more frequently to colour shops, but at the very end of the eighteenth century colourmen began to write books themselves. The first manufacturing artists' colourman to have published a literary work appears to have been Constant de Massoul, who wrote *A Treatise on the Art of Painting and the Composition of Colours.* It was originally written in French and published in France, but in the English edition of 1797 the author states that the colours listed were on sale at his factory in New Bond Street, London. The long list of colours includes synonymous names which are explained in the text, but it contains such a strange mixture of modern and traditional colours that it is doubtful if all of them were really supplied commercially. In this respect *A Treatise on Ackermann's Superfine Water Colours,* which was published in 1801 by Rudolph Ackermann, the printseller and colourman, is extremely useful, as, although the first part on the sources and nature of various pigments is acknowledged to have come from de Massoul, there is also a list of colours actually sold by Ackermann. The total number is only sixty-eight, which, when compared with de Massoul's list of eighty, reveals the omission of some traditional colours, such as blues made from turnsole and cornflowers. It can be regarded as presenting a much more reliable list of the colours in common use at the time. An outstanding feature of both books is the lack of secrecy about the composition of most of the colours supplied commercially. Although they explain methods for preparing various colours, the authors were apparently confident that few people would attempt to do so because, as Ackermann says, 'The time, the trouble, the expense attending their preparation will never compensate the small saving gained by it.'

Possibly because full colour lists were available elsewhere, many books published in the early nineteenth century do not contain full details of the numerous colours available but provide a short list of recommended colours instead. Such books are of limited use for the history of pigments, although they sometimes indicate when new colours were accepted in general use. Early nineteenth-century books for amateurs and beginners are almost too numerous to mention; they include *Hints to young Practitioners* by J. W. Alston, 1804, *Instructions for Drawing and Colouring Landscapes* by Edward Dayes, published posthumously in 1805, *An Essay . . . on Colours* by M. Gartside, 1805, *The Seasons . . . instructions for drawing and painting flowers* by P. C. Henderson, 1806, *Bryant's Treatise on the Use of Indian Ink and Colours,* 1808, *The Art of Painting on Velvet* by the Reverend T. Towne, 1811, *The Elements of Drawing* by George Hamilton, 1812, *A Compendium . . . of Painting* by Richard Dagley, first published in 1818 and reprinted in 1822, *The Young Artist's Guide* by W. H. Edwards, 1820, and *The Practice of . . . Painting Landscape* by Francis Nicholson, second edition 1823. Some of the books have plates coloured in water colours, but none has named colour samples. Some other books may be ignored, because they contain hardly anything actually written during the nineteenth century. The responsibility for publishing out-of-date works must rest with Richard Holmes Laurie, a map and printseller of Fleet Street, who reissued one of the later editions of *The Art of Drawing,* 1731, as *The Art of Painting in Water-Colours,* Hoofnail's book as *The Painter's Companion,* John Smith's book on oil painting under the title *Smith's Art of House-Painting,* and also John Payne's book on miniature painting under the author's name and with the original title.

In general, late eighteenth-century and early nineteenth-century books which contain named colour samples are more useful than any of the others. The professional painter, Julius Caesar Ibbetson (1759–1817), produced two books with colour guides. The lesser known of the two is

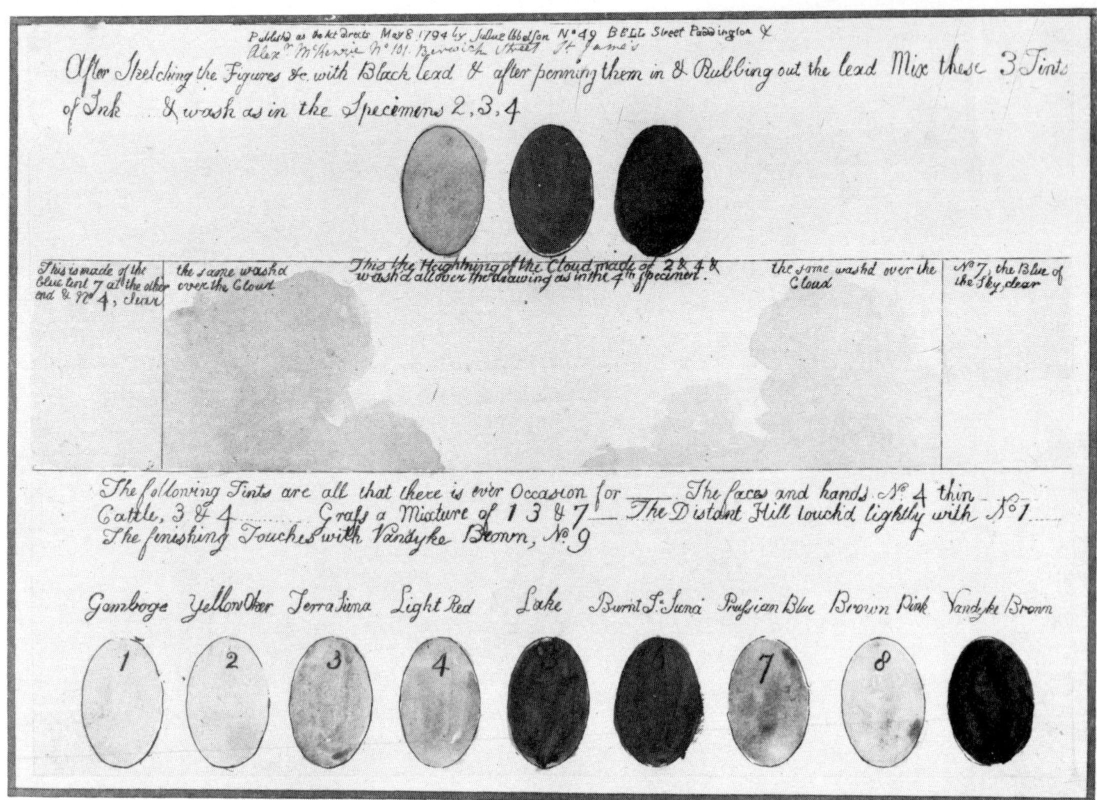

After Sketching the Figures &c. with Black lead & after penning them in & Rubbing out the lead Mix these 3 Tints of Ink & wash as in the Specimens 2, 3, 4
This is made of the Blue tint 7 at the other end & No 4, clear
the same washd over the Cloud
This the Heightning of the Cloud made of 2 & 4 & wash'd all over the drawing as in the 4th Specimen.
the same washd over the Cloud
No 7, the Blue of the Sky clear
The following Tints are all that there is ever Occasion for — The faces and hands No 4 thin — Cattle, 3 & 4 — Grass a Mixture of 1 3 & 7 — The Distant Hill touch'd lightly with No 7 — The finishing Touches with Vandyke Brown, No 9
Gamboge
Yellow Oker
Terra Sina
Light Red
Lake
Burnt Sⁿ Sina
Prussian Blue
Brown Pink
Vandyke Brown
1
2
3
4
7
8

Fig. 9. A limited water-colour palette recommended by Ibbetson. Four of the nine colours were fugitive. (J. C. Ibbetson, *Process of Tinted Drawing*.) (Reproduced by courtesy of the Trustees of the British Museum.)

his *Process of Tinted Drawing,* which is undated but was probably printed in 1794–95. It consists of nine prints, including a colour chart, bound together as a book. Apart from the notes and colour names printed on the chart, which includes nine water-colour washes and some sample mixtures for sky tints, there is no text, and, probably for that reason, the book is seldom mentioned in connection with Ibbetson. *An Accidence or Gamut of Painting,* 1803, is known for Ibbetson's condemnation of picture restorers and his recommendation of gumption, an oil medium based on mastic. It also contains a number of named colour specimens and sample mixtures.

Some early nineteenth-century books contain only a few colour washes; amongst these are *A Practical Essay on the Art of Colouring* by J. H. Clark, 1807, and *A Treatise on Landscape Painting* by the painter David Cox (1783–1859), who states that it was produced at the request of his pupils. Named samples are also supplied in *Introductory Lessons . . . in Landscape,* privately printed by James Roberts in 1809, and in *A Practical Essay on Flower Painting* by Edward Pretty, 1810. *J. Varley's List of Colours,* 1816, consists of a single sheet with water-colour washes and a short text with comments on each of the colours. John Varley (1778–1842) was a professional painter who taught a number of subsequently well-known artists. The colour chart is said to have been produced for his pupils, and it is valuable in showing the palette used by professionals instead of the colours likely to have been used by amateurs for whom most of the other books were intended. The *List of Colours* was reissued several times during the early nineteenth century. T. H. Fielding, another teacher of painting, was the author of the *Index of Mixed Tints,* published in 1830, which has twenty-eight named colour samples and additional washes showing the various mixtures obtainable.

In many ways, the most useful literary source concerning pigments rather than painting of the period is Field's *Chromatography,* published in 1835. George Field (1777?–1854) was known also as the author of *Chromatics,* a book on colour theory, and, perhaps more important to professional painters, as a colour-maker. (The distinction between a colour-maker and a colourman is that the first manufactures the pigments whereas the second makes artists' colours, preparing pigment and

Fig. 10. George Field (1777?–1854): mezzotint by Lucas after a portrait by Constable. (Reproduced by kind permission of Messrs Winsor and Newton Limited.)

medium together without necessarily manufacturing the pigments.) Very little can be learned about Field's manufacturing activities from his books; however, the ten manuscript notebooks that have survived contain much important information on that subject. In addition, the notebooks that were compiled over a period of years, largely 1804–25, provide the basis of the remarks on the properties of pigments that are included in the printed work, *Chromatography,* making it clear that Field had tested most if not all the pigments, and that he had a close relationship with a number of British and American painters.[24] It seems fairly certain that colourmen purchased pigments from Field, although nineteenth-century references to Field suggest that he also dealt directly with some artists. The first part of *Chromatography* is concerned with colour theory, followed by a section in which the nature and composition of individual pigments are discussed. The last part contains some comments on oils, varnishes and picture cleaning. The frontispiece illustrating colour theory includes some small water-colour samples which are not named on the plate but are identified in the text. From an historical point of view, an important feature of the first edition is the inclusion of a large number of pigments, no matter how obscure, so that the book fills the gaps left by most of the early nineteenth-century books on painting. The first edition is written in a more literary style than the later editions of 1869 and 1885 (edited by T. W. Salter and J. Scott Taylor respectively) which include modern colours and omit the poetical quotations and the comments on little-known or obsolete pigments.

The description of the contents of literary sources dating from 1600–1835 will have shown quite clearly that very few contain details of manufacture. The deficiency cannot be made up by reference to books on colour chemistry, for next to none were written at that time, although *The Art of Painting in Oil* by Mérimée comes into that category and only just falls outside the period. The original edition was published in French in 1830, and an English translation appeared in 1839. Early scientific papers and works on technology, which are discussed in the following chapter, provide a certain amount of information on isolated subjects, but they do not supply evidence of continuous development over a long period. The value of the literary sources lies in the fact that they supply a continuous commentary on the pigments used, their preparation as artists' colours, and the purpose for which they were best suited. In addition, they provide much useful information concerning English colour names. Some previous writers have ignored changes in meaning and have, therefore, misinterpreted some names, so, for that reason, considerable attention is given to nomenclature in the following chapters on pigments. Field's *Chromatography* provides a suitable close to the survey of literary works, not merely by virtue of the fact that it lists the range of colours available in 1835, but because in it the author shows a responsible concern about the durability of colours which largely gave rise to later editions of the book by others and so prepared the way for scientific works on colour chemistry in the late nineteenth century.

3

Sources for the History of the Colour Trade

Pioneer historians of artists' colours (Eastlake and Mrs Merrifield, for example) largely based their work on literary sources and records of workshop practice. Such documents throw some light on the raw materials used; they may indicate whether a pigment was natural or manufactured and whether it was a native or a foreign product, but they do so only in general terms. Much more evidence is to be found in other documentary sources, such as the public records and records of scientific societies. The relevant public records form an important group of documents, which, since they have not previously been studied with reference to pigments, will be described in detail.

First of all, there is the question of native and imported raw materials. This may not appear to present much of a problem, because many colour names suggest foreign origin. Ultramarine, perhaps the most evocative of traditional colour names, literally means the colour which comes from overseas. Cologne earth is another name with obvious foreign associations. However, there are borderline cases; for example, realgar, a name derived from the Arabic phrase meaning powder of the cave, indicates a mineral of foreign origin, although there is a possibility that realgar used in seventeenth-century England might have been English, either natural or manufactured. Other names, such as Spanish brown, have seemingly obvious foreign associations which upon investigation appear purely traditional and quite irrelevant. Problems such as these can be resolved by reference to public records, those which supply evidence on imports being the customs records and documents of the seventeenth-century chartered trading companies.

The customs records may be divided into the customs ledgers and the port books, of which the former are the more important for the purposes of a general survey, although they are, unfortunately, available only from the very end of the seventeenth century. Reference was made to a selection of ledgers between 1700 and 1830, and, on that account, the discussion of imports in the following chapters is with very few exceptions restricted to the eighteenth and early nineteenth centuries. Examination of a selection of documents does not provide complete evidence although it gives an overall view, and, in this case, it provided information concerning the most important imports and their places of origin, commodities such as madder and indigo appearing in all the ledgers examined.[25]

A certain amount of care is required in the interpretation of customs records, as one must not automatically assume that the country of origin listed was the actual source of any material. For example, it is reasonable to assume that madder imported from Holland was actually Dutch, but one must recognise that cochineal imported from Holland was of Latin American origin. This problem applies particularly to goods imported from maritime countries, although during war, particularly with France, goods were brought in from non-combatant countries instead of the country of origin. Another slightly misleading feature is the fact that any ship which had travelled from the East round the Cape of Good Hope was listed as coming from the East Indies, a vague term which could mean anything from the sub-continent of India to China. Thus it is that carbon black ink came to be known as Indian ink, although it was imported from China.

Fig. 11. Chinese stick ink, so-called Indian ink. Some examples are very decorative, some gilded. They must be rubbed down with water on ground glass or in a saucer ready for use.

The customs ledgers are ostensibly written in English, but it is often possible to trace the survival of an old form of spelling in official documents, the customs records included. They contain a certain number of references to pigments by Latin names: Armenian bole appears as *bolus Armeniacus,* orpiment as *auripigmentum,* madder as *rubia tinctorum.* In all cases the same pigment appears under its English name in some ledgers and in Latin in others, so that one must be prepared to encounter both forms. Italian occurs in the names *terra verde* and *terra umbra,* but foreign articles are generally listed in English, subject of course to the clerk's capacity to transcribe outlandish words, such as gamboge, gum senegal and gum tragacanth, all of which gave rise to difficulty and innumerable spelling variations. A survival of ancient tradition is to be found in the identification of some pigments as drugs. Mention has already been made of the fact that *pigmentum* originally meant a drug, and it is interesting to find that as late as 1700 madder roots, orpiment and verdigris which were imported from Holland were bracketed together as drugs, whereas lead white from the same source was listed separately. In comparison with the customs ledgers, there is little to be said about the port books, which are included amongst the documents appertaining to the Exchequer. A large number of seventeenth-century port books have survived, but it is impracticable to use them to fill the gap left by the absence of annual summary customs ledgers for the seventeenth century, as the sheer bulk of mixed information contained in the daily records of the entire cargo of every ship precludes their use for a survey of a general nature.

In some ways, there is a similar difficulty in extracting information concerning a few commodities from records of the chartered trading companies, the most important for the subject of pigments being the Levant Company and the East India Company. Reference to printed histories of the former showed that there was little likelihood of obtaining useful information from original sources, since very few letters from the merchants at various trading stations have survived from the early seventeenth century, an important period which covers the declining years of the overland route from the East. On the other hand, it was possible to make use of the records of the East India Company because the earliest letters from the Company's merchants have been transcribed and printed, and, in addition, summaries of the early Court Minutes are available in printed form, thus allowing easy reference to the original documents in the India Office Records.[26] The general information contained in the Company's early seventeenth-century records hardly suggests that the Company's servants in the East were, in competition with the Dutch and Portuguese, playing an important part in a commercial revolution. The records contain numerous recriminations; the factors complained that the Company officials sent them unsaleable goods to trade, while the officers grumbled about poorly packed and loaded cargoes. Letters home are full of complaints about the iniquities of the Dutch who dominated the islands of the East Indies and therefore had the richest trade, the Indians who, amongst other things, sold wet indigo, and the Arabs who stole shipments which were off-loaded for the land passage between the Red Sea and the Mediterranean, a route which the Company used during the early years in addition to the long sea route round Africa. Yet despite all the difficulties of which the British were so conscious, the Mediterranean countries were hard hit by the changing pattern of trade. Their point of view is presented in the Calendars of State Papers, which contain summaries of letters written by the Venetian ambassador in London and decisions made by the governing body in Venice. The original documents are preserved at Venice, but they are of such value to British historians that the contents have been incorporated in printed public records of the early seventeenth century.[27] Although they do not include specific remarks on the subject of pigments, they throw a considerable amount of light on the commercial success of the British and the difficulties of the Venetians. The evidence suggests that, whereas the Venice ceruse and Venice lake mentioned in English literary sources of the sixteenth century may well have been Venetian in origin, there is reason to regard as spurious references to the same commodities in similar sources of the seventeenth century, for the British had the necessary raw materials to make their pigments themselves. The question remains as to whether or not they had the necessary knowledge.

For the home trade, the public records which supply most evidence are those concerning patents for invention, grants made to manufacturers giving them sole right for a limited period to use a certain process or make a particular article. The nature of the evidence varies between patents dating from the sixteenth and seventeenth centuries and those from the eighteenth century and later, but, before discussing the differences between them, a brief description of the documents is included, for many exist in both manuscript and printed form.

Letters patent were grants made by the Crown to one or more persons, that is, they were declarations in the form of open letters as opposed to private communications. One copy was sent to the patentee and a copy was entered on the patent rolls, but for many years there was no provision for a subject index of patents for invention nor in fact any record which distinguished them from the other letters patent on the rolls. It was not until the middle of the nineteenth century that the numbering of patents for invention was undertaken, at which time a series of blue books was printed, the series including a book for every patent which received a number, each book containing a transcript of the terms of the grant or, in the case of patents of later date, a transcript of the patent specification. Unfortunately, the early part of the series is defective because it was compiled from entries in the docket books of the Clerk of the Letters Patent instead of from the actual patent rolls, with the result that there are no entries prior to 1617. The defect was noticed fairly soon, and various officials at the Patent Office have undertaken the necessary research which makes it possible to trace patents for invention which are not included in the blue books.[28]

Towards the end of the sixteenth century letters patent for invention were granted with increasing frequency, largely because they were regarded as a useful means of encouraging national industry and not, at that period anyway, as a non-parliamentary source of revenue. At that time the grant was not necessarily made for an invention in the modern sense but frequently for the introduction of a manufacturing process which had previously been practised overseas. Then and throughout the seventeenth century the patentee was under no obligation to submit a specification nor offer an explanation of his process or invention, and, for that reason, one cannot expect to gain much technological information from early patent literature. The whole justification for a grant rested on the patentee's ability to conduct the industry in which he had a monopoly; consequently, early patents supply evidence that certain industries were practised for a time at least, even though some patentees were unsuccessful in continuing manufacture for the whole period allotted to them. Seventeenth-century industry suffered as a result of the system, the shortcomings of which are illustrated in Appendix Three by a description of the difficulties encountered by some early industrialists engaged in smalt-making and the preparation of indigo.

The development of patent specifications must be taken into account when considering evidence supplied by eighteenth-century patent literature, for at that time it became first customary and then obligatory to enter a specification, so that the validity of a patent rested upon the submission of a specification within a certain period and, also, upon the novelty of invention in the modern sense. There was no compulsion to put the patented process into operation; consequently, eighteenth-century patent documents cannot be treated as evidence of manufacture and they are not mentioned in the following chapters unless there is other evidence that the process was put into operation. Patent specifications dating from the second half of the eighteenth century are often extremely verbose and ambiguous. It is necessary to take into account the fact that as legal documents they contain much repetition which should ensure absolute clarity, but as documents concerned with science and technology they suffer from complexity attendant upon the inadequacies of eighteenth-century chemical terminology. One is led to suspect that a patentee sometimes deliberately made a specification obscure so that his contemporaries would find it difficult to pirate the invention, but it also seems possible that some manufacturers found it difficult to explain a process in precise terms. The troubles faced by a manufacturer of patent yellow and an importer of quercitron are discussed in later chapters. In both instances the patents are extremely valuable, because they supply the earliest possible date for the use of those pigments in England. Some early seventeenth-century patents are equally useful but in a different way, for they supply information about artisans and industry at a time when both were totally ignored by writers and scientists.

An indication of the general attitude of men of learning in 1605 is given in the following passage written by Francis Bacon:[29]

> For History of Nature Wrought or Mechanical, I find some collections made of agriculture, and likewise of manual arts; but commonly with rejection of experiments familiar and vulgar. For it is esteemed a kind of dishonour unto learning to descend to inquiry or meditation upon matters mechanical, except they be such as may be thought secrets, rarities and special subtilties.

At that time the author was almost isolated in his condemnation of such an attitude as 'vain and supercilious arrogancy' and in his view that the study of trades would lead to practical improvements and that general experiments could bring about the establishment of universal truth or scientific principles. But Bacon's comments took root, and it is generally accepted that his writing was a formative influence in the creation of the group which gathered first at Oxford in 1645 and emerged later in

London, receiving the charter of incorporation as the Royal Society in 1662.

The influence of Bacon may be detected in some of the activities of the Royal Society in its early years. For example, his recommendation of empirical method is reflected in the variety of experiments undertaken of which many, if not quite in the same category as Defoe's sunbeams from cucumbers, were certainly somewhat bizarre. Also amongst early activities was an investigation into various manufacturing processes for the purpose of producing a series of 'histories of trade'. The Fellows formed the resolution to produce the series in a spirit of enthusiasm which was regrettably unmatched by application, but the episode is of some interest because the subject of artists' colours was to have been included. The idea undoubtedly originated with John Evelyn, the diarist, who in 1661 prepared a classified list of trades which still survives.[30] He divided occupations into eight classes, the first of which, 'useful and purely mechanical', included colour-maker, ceruse-maker, brush-maker, painter–stainer and red-lead maker, all of which were superior to class two 'meane and less honourable' and class three 'servile', but not in the same class as 'polite and more liberall'. Fine art (oil painting, miniature, mosaic and fresco) was not included with the polite arts but amongst 'curious' arts, so that artists were not promoted to the rank that Hilliard would have liked but were nevertheless separated from decorators, the role of most painter–stainers. Class eight, 'exotic and very rare seacrets', included varnishes, Chinese ink and ultramarine, lake and ceruse. The inclusion of the last item somewhat spoils the list because ceruse is also mentioned in class one and there was nothing secretive about its manufacture, but the remainder of class eight gives some idea of the commodities which were regarded as extraordinary. Some accounts, or histories as they were called, of different trades were written, and, in 1666, a Fellow presented a paper on illuminating and promised to exert his influence to obtain an account of enamelling from Petitot. The idea was received enthusiastically, and a committee was formed to enquire into different painting techniques in collaboration with several artists, amongst whom Lely, Cooper and Streeter were named. A year later nothing had come of the project, although all three artists were reported as having agreed to take part.

Their willingness is worth emphasising because the refusal of another, Alexander Marshall, has received more publicity, as his letter refusing to divulge secrets (which almost certainly concerned lake-making) is printed in Birch's *The History of the Royal Society,* a printed version of the Society's journal from 1660–87. It is true that Marshall, a water-colour painter, wrote '. . . they are pretty secrets, but known, they are nothing. Several have been at me to know, how; as if they were but trifles, and not worth secrecy. To part with them as yet I desire to be excused.'[31] However, too much should not be made of his attitude, for reference has already been made to artists of an earlier period who passed information to de Mayerne, and there is no reason to suppose that Lely would not have done the same for the committee if the members had been more assiduous, for one of them, Robert Hooke, had previously been his apprentice.[32] The failure of the scheme to produce a history of artists' colours was probably the fault of the committee rather than the artists, as Hooke was too busy with his work as staff scientist at the Royal Society, and Evelyn lost interest in trades because of 'many subjections, which I cannot support, of conversing with mechanical capricious persons'.[33] There can be no doubt that the members originally had the best of intentions, but after the first decade of the Society's existence the emphasis of its activities was increasingly directed towards pure science rather than technology and nothing more was heard of the histories of trade. After the election of Newton in 1671, colour and light received far more attention than pigmentary colours.

Although general interest in artists' colours faded, some Fellows had already made or were to make individual contributions on the subject. Robert Boyle's work, *Experiments and Considerations touching Colours,* has been mentioned earlier with reference to its regrettable influence on

Fig. 12. Robert Boyle (1627–1691): copy after a portrait by Johann Kerseboom, *c.* 1689–1690 (Reproduced by courtesy of National Portrait Gallery).

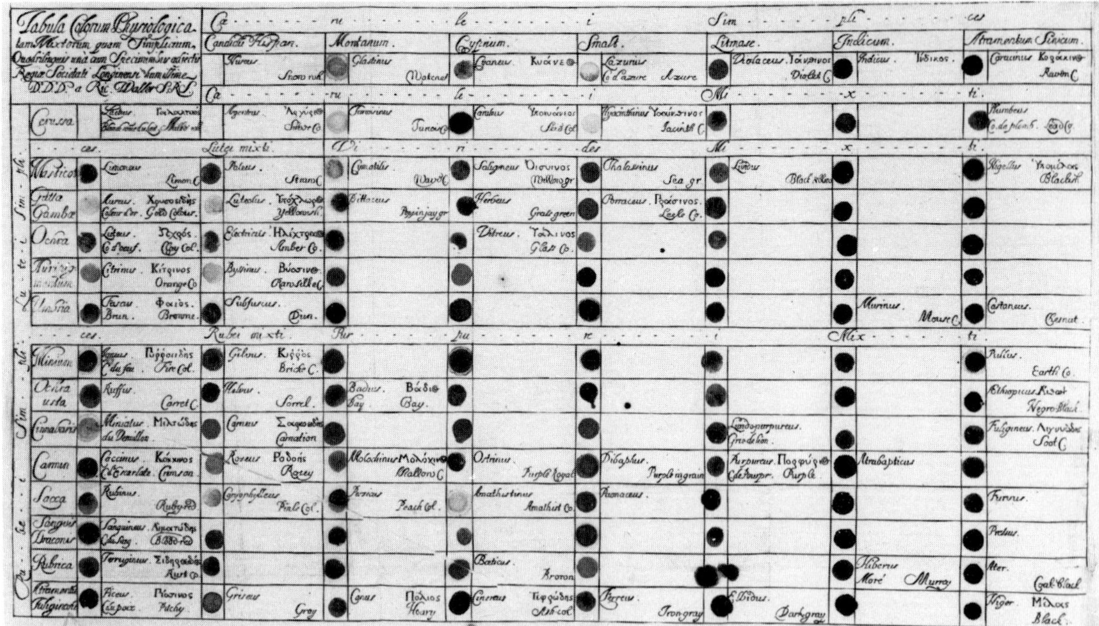

Fig. 13. Page of colour samples accompanying Waller's paper in *Philosophical Transactions*, xvi (1686). Blues from light to dark are arranged left to right on the top row; white, yellows, reds, browns and black are arranged from top to bottom in the left column. The remaining samples are mixtures of these colours. (Reproduced by courtesy of the University of London Library.)

eighteenth-century books on painting, owing to the author's claims for the artistic value of dyes which most contemporary painters would have rejected as nonsense. Nevertheless, Boyle, who did not mind conversing with artisans, appears to have been acquainted with some aspects of colour manufacture and both his books on colours and an earlier work, *On the Unsuccessfulness of Experiments,* 1661, contain a few interesting comments. For practical purposes, however, Christopher Merret's edition of *The Art of Glass, wherein are shown the wayes to make and colour Glass, Pastes, Enamels, Lakes, and other Curiosities,* 1662, is much more useful. The book contains a translation of *L'Arte Vetraria* by Antonio Neri, with additional notes by Merret. Although the author of the Italian original died in 1614 and much of the contents of the English book were, therefore, traditional and not the products of seventeenth-century research, the existence of a printed account in English of lake-making methods gives the lie to the common belief that the English had to import red lake pigments because they were ignorant of the proper way to make them. The main subject, glass-making, is better treated and much expanded in Kunckel's illustrated *Ars Vitraria Experimentalis,* 1679, a German translation of Neri's and Merret's work with original material by Kunckel, an alchemist who at one time worked in Saxony and was later minister of mines to the king of Sweden.

A few papers printed in the Royal Society's *Philosophical Transactions* in the seventeenth century concern pigments or artists' colours. One of the most interesting is 'A Catalogue of simple and mixt Colours, with a

Specimen of each Colour prefixt to its proper Name' by Richard Waller.[34] As the title suggests, a chart is included with named colour samples and mixtures. Waller, who hoped to establish a standard for the decription of various hues and intermediate colours, included short comments on a selection of pigments, some of which are erroneous because he referred to classical authors without taking into account the changed meaning of some colour names. A number of his remarks are interesting, however, particularly the striking observation that massicot is made from tin. When Waller's comments are mentioned in the following chapters, the reference is to his paper in *Philosophical Transactions.* Papers by other Fellows of the Royal Society are also referred to below, but most of them are devoted to one particular pigment or raw material and so require no general description here. Amongst the most interesting is a series of papers concerning the nature of kermes and cochineal, some of them by Martin Lister, the zoologist who was also a friend of the painter Henry Gyles.

The documentary resources of the Royal Society proved to be most useful for the seventeenth century, for very few papers relating to pigments are to be found after 1700.

After the Fellows of the Royal Society became less interested in trades and manufacture, a long interval elapsed before members of another body took up the subject. Consequently, there are few documentary sources touching on colour manufacture which date from the first half of the eighteenth century. Some books from that period, Shaw's *Chemical Lectures,* 1733, for example, refer to substances which were used as pigments either then or later, but rarely discuss pigments as such. Thus, there is a half-century which supplies few documentary sources concerning the history of artists' colours, for the lack of source material on manufacture is comparable with a similar scarcity of literary sources on painting technique.

The activities of the Royal Society of London and the Académie des Sciences in France served as an example to others, and in the middle of the eighteenth century a number of scientific societies were established in other European countries, including Germany, Holland, Denmark, Sweden and Italy. As part of the same general movement, the Society for the Encouragement of Arts, Manufactures and Commerce (now the Royal Society of Arts) was founded in England in 1754, and interest in manufacture was revived. An important member during the early years of the Society was Robert Dossie, a chemist who acquired an international reputation with his book *The elaboratory laid open,* which was published in the same year as *The Handmaid to the Arts.* The aims of the Society of Arts were expressed when he dedicated the latter work to the Society:

> The furnishing means of establishing and improving useful arts, especially those which relate to considerable manufactures, and the creating incitements and motives to the exercise of those means, are to a country that owes it riches, power, and even domestic security to

> commerce, of the greatest concern and moment; and it is more peculiarly meritorious in those, who, in a private capacity, exert their utmost endeavours on these accounts, as such pursuits seem to take up a very little share of the regard of the public here, at a time, when all the neighbouring governments (especially that of our rival France) make them a principal object of their attention and care.

It is clear from the contents of *The Handmaid to the Arts* that Dossie had succeeded in obtaining much knowledge of manufacturing methods, but how he set about gathering the information has never been revealed.[35] Another of Dossie's works, *Memoirs of Agriculture and other Oeconomical Arts,* three volumes 1768–82, is also of interest, because it gives an account of the ways in which the Society of Arts tried to encourage industry during the years before the Society published an account of its proceedings. The Society offered prizes or premiums to those who could supply evidence of having produced a specified quantity of a certain material, such as madder roots or smalt. As the campaign to encourage any product progressed, the number of claims for premiums increased, and, when the campaign was judged to have achieved its object, no further premiums were offered. In addition, the Society awarded prizes for inventions or improved industrial methods with the proviso that no prize should be given to a patentee or anyone who intended to apply for a patent. Members of the Society were absolutely firm on this point, and their disapproval of the patent system was recorded in a volume of the *Transactions*:[36]

> Several decisions lately made in the Courts of Justice, show upon what futile grounds many of the Patents stand which have been granted, and that, instead of producing any profit to the Patentee, they have involved him in great expense and trouble; whereas the rewards of the Society are granted free of every expense, and the inventor's name is recorded and handed down to posterity.

The Society doubtless did much good work in giving publicity to various manufacturing processes, and its detailed records now supply evidence on some aspects of particular trades. When an award was made, details were published in the *Transactions,* but the manuscript minutes of the committees which considered each award contain additional information. For example, the minutes of the committee of chemistry and mechanics show that in 1815 George Field was conducting lake-making from his home at Heath Cottage, Hounslow Heath, where several of the members went to see the different pieces of apparatus in use (*Figure 53*). Another colour-maker who received an award was James Rawlinson of Derby, whose improved grinding mill (see *Figure 14*) was commended in 1804 when Middleton the London colourman (often mentioned in books on painting and therefore probably the best-known colourman at that time) stated that he intended to put one of the machines into immediate use, 'as the demand for the Colours it is calculated to grind is very considerable'. As it happens, the records of the

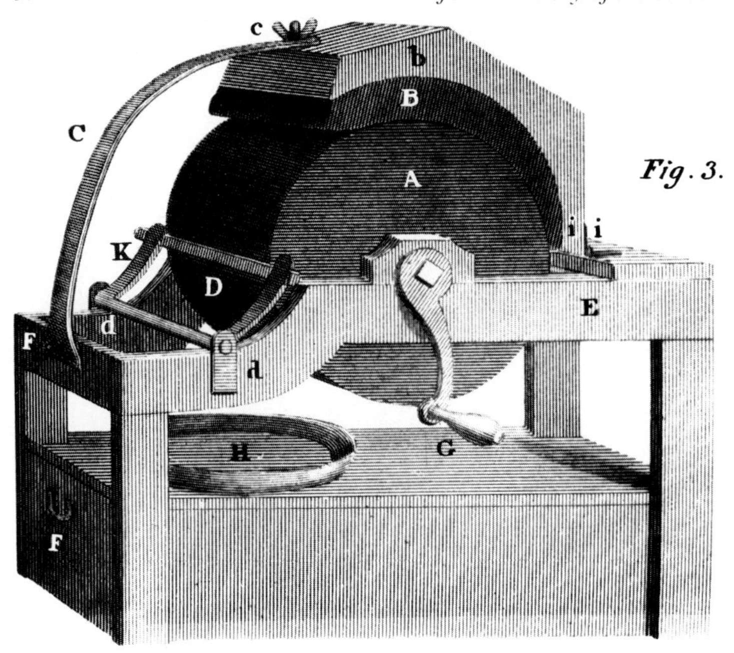

Fig. 14. Rawlinson's hand-operated, single-roll grinding mill, which was recommended by the Royal Society of Arts in 1804: A is the marble roller and B the concave muller covering one-third of the roller: C is a piece of iron to keep the muller steady (the pressure of the iron could be increased by tightening the fly-nut *c*); the spring D took off the colour when the action of the roller was reversed, and the paint was collected in the tray H. The muller was designed to be turned back on pinion *ii*, so that the mill could be cleaned with curriers' shavings which were kept in a drawer below the mill (Royal Society of Arts, *Transactions*, xxii, 1804). (Reproduced by courtesy of the University of London Library.)

Society of Arts are more useful on the subject of artists' colours than pigments, and an examination of the index to the *Transactions* between 1783, when it was first printed, and 1835 showed that it is a fairly good source on painting and painting equipment, although some subjects reflect the late eighteenth-century and early nineteenth-century pre-occupation with the 'secrets' of the Old Masters.

The problem of secrecy, not necessarily in painting technique but in industry or any other activity, is two-sided; to those who are in ignorance anything may be a secret, but that does not mean to say that the knowledge is deliberately withheld. The early records of both Royal Societies, of London and of Arts, reveal the measure of ignorance on some matters. For example, both societies enquired into the nature and manufacture of saffer and smalt as if it were something rare, but it took neither of them very long to discover others who knew about its manufacture. Despite the good intentions of both societies, a certain lack of communication between industrialists and scientists persisted, largely, one may suppose, because the artisans were too busy with the day-to-day work of earning a living to take part in the activities of learned societies unless lured by the offer of a premium or prize. It is an accepted fact that the early part of the period known as the industrial revolution saw the introduction of new ideas and machines in agriculture more than industry, a phenomenon which may be accounted for because many from the educated upper class were also landowners. The slowness of the artisan to adopt improvements may have resulted from lack of education, which in turn led to a lack of awareness concerning matters outside his immediate environment. A similar theory was suggested when the formation of the Royal Institution was proposed in 1799:[37]

> There are no two classes of men in society, that are more distinct, or that are separated from each other by a more marked line, than philosophers and those who are engaged in arts and manufactures. The distance of their stations—the difference of their education, and of their habits—the marked difference of the objects of their pursuits in life—all tend to keep them at a distance from each other, and to prevent all connection and intercourse between them.

Whereas the Society of Arts had been formed to stimulate agriculture and industry and to help the nation's economy, the Royal Institution was a humanitarian body, founded to disseminate knowledge of new inventions and to educate artisans to accept and make use of them. A repository or permanent exhibition of new machinery was planned as a visual aid to education following the idea that 'something *visible* and *tangible* is necessary to fix the attention'. Lectures were also planned, and everything possible (except the award of monetary prizes) was to be done to encourage the adoption of new ideas. The humanitarian aims of the society are evident in proposed subjects upon which the artisan should be instructed—the need for proper ventilation, for example—but notices of scientific research were printed in the *Quarterly Journal of Science and the Arts,* which appeared some years after the foundation of the institution.

By the beginning of the nineteenth century it was common for scientists to publish details of their research in scientific journals, and their papers were frequently translated so that there was no lack of communication between the scientists of one country and another. In addition, there were a few journals, including the *Quarterly Journal* of the Royal Institution and the *Repertory of Arts* (later the *Repertory of Patents of Invention*), which specialised in relaying information on the technological application of new discoveries. The latter included details of some inventions which received awards from the Society of Arts and of some patented processes. However, books and papers on chemistry exceeded those on technology, and, although a small amount of information concerning pigments is to be found in books such as *Chemical Essays* by Richard Watson, an eighteenth-century Fellow of the Royal Society, and another with the same title by Samuel Parkes published in the early nineteenth century, it was not until later that books on manufacture were written. Amongst them one may cite Andrew Ure's *A Dictionary of Arts, Manufactures and Mines,* 1839, written by a doctor of medicine who was later a professor of chemistry and physics, and *Days at the Factories,* 1843, and *Textile Manufactures,* 1844, by George Dodd, who specialised in writing about industry.

Then and throughout the whole period under discussion it was generally the branches of colour-making applicable to decorating as well as fine art which received most attention, as for instance the manufacture of white and red lead and verdigris. Other pigments were produced either as a by-product in another industry, just as copper carbonates were made by silver refiners, or they were manufactured in relatively small quantities for artistic purposes, in which case the manufacturer was likely to operate a home industry. It seems that the colourmen drew on many suppliers, and their own manufacturing activities were too limited for Dodd to write a detailed account although he did not ignore them completely:[38]

> There are several tradesmen, denominated 'artists' colourmen', who, in the preparation of primed canvas, card-board, oil-colours in bladders, water-colours in cakes, brushes, pencils, palettes, easels, &c., carry on a tolerably busy circle of operations. But these various articles are rather a collection of component parts, procured from workmen in various quarters, and brought into a saleable form, than the result of a uniform system of manufacturing.

The manuscript records of two colour-makers in England in the late eighteenth and early nineteenth centuries have been used for this study;

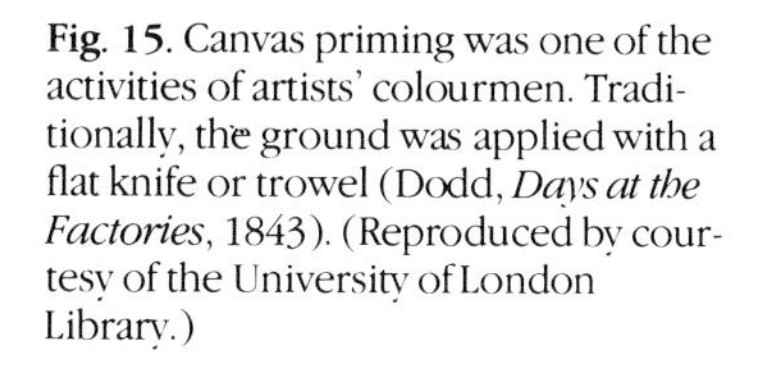

Fig. 15. Canvas priming was one of the activities of artists' colourmen. Traditionally, the ground was applied with a flat knife or trowel (Dodd, *Days at the Factories*, 1843). (Reproduced by courtesy of the University of London Library.)

they are experimental notebooks, formula books and stocklists of Berger (now Berger, Jenson & Nicholson Ltd) and the manuscript notebooks of George Field. It is interesting that names of many of the British artists' colourmen who were well-known in the nineteenth century appear in these documents, making it quite clear that both Berger and Field had connections with London colourmen such as Ackermann, Blackman, Brandram, Middleton, Newman, Reeves, Rowney and Strachan. Berger supplied most of them, as is clear from his very early nineteenth-century stockbooks. At the same period, quite early in his career, Field supplied some of them and purchased pigments from others for test purposes.

The early printed catalogues of the artists' colourmen, Winsor & Newton Ltd, were also used in this study of artists' pigments. The complexity of the colour trade is apparent in the very lengthy colour lists, with seldom fewer than fifty colours for sale, and often many more, with colour names that could indicate a certain pigment at one period and a totally different pigment some years later.

Owing to the complexity of the colours available and the diversity of the colour trade itself, which involved merchants, colour manufacturers and artists' colourmen, the subject is best studied colour by colour.

4

Inorganic Blues

Natural Ultramarine: $3Na_2O.3Al_2O_3.6SiO_2.2Na_2S$

Ultramarine was for centuries the most highly prized of all traditional pigments, not only on account of its intrinsic value but also for its durability and excellent colour. In some sixteenth-century sources the same pigment is called *azure,* a name which is associated with its hue and that of lapis lazuli, the blue stone from which it is made. Lapis lazuli is a mineral of the sodalite group which occurs in several parts of the world, one of the best known being the mine in the Kokcha valley in Afghanistan, believed to have been the source of most of the lapis lazuli used in Europe during the Middle Ages and the early modern period. The mineral occurs in mountainous regions, often remote and almost inaccessible, and the deposit in Afghanistan is no exception. Europeans knew little about it, and no description of the mine was written in English until 1837 at a time when it was unworked.[39] Difficult access to the mine must have made even the start of the long journey to Europe extremely arduous, whether the lapis lazuli was carried overland to the Mediterranean or south to India.[40]

Painters were not primarily concerned with the original source of the stone, for the place of purchase was of greater importance. Hilliard refers to the colour as *ultermaryne of Venice,* which reflects the supremacy of Venice in Mediterranean trade until 1600, but his reference marks the end of an era, for Norgate states that lapis lazuli is obtainable from the Aleppo merchants. His remark was based on first-hand knowledge, as he had visited the Levant and knew of the activities of the British merchants of the Levant Company who had a permanent trading station at Aleppo. Development of British trade in the eastern Mediterranean in the early seventeenth century undoubtedly ensured that the British obtained at least some of the lapis lazuli then obtainable, just as later development of trade with India ensured that they secured much that was exported through India in the eighteenth century.

There is good reason to suppose that lapis lazuli was available in England during the first half of the seventeenth century for those painters who could afford it. Detailed instructions for the preparation of ultramarine exist in several documentary sources. Some begin with instructions concerning the selection and purchase of the raw material, recommending the best lapis lazuli as being deep blue and as free as possible from streaks of grey or gold-coloured flecks of pyrites, variations which were undesirable in colour-making but at least showed that the stone was genuine. Both lapis lazuli and the pigment made from it could be tested by heat, because, whereas the genuine article was unaffected and remained blue, azurite which might be substituted for the stone would turn black and a substitute pigment was also likely to change colour.[41] Heating was, in any case, the first step in the treatment of lapis lazuli as, when it was afterwards plunged into a cold liquid, it was liable to fracture and could then be broken up more easily. Then the blue mineral (now known as lazurite) had to be separated from the rest of the stone. Some instructions for a quick method exist, as, for example, in de Mayerne's note that the stone can be crushed, vinegar poured on and the mixture heated so that the blue colour will eventually be drawn out into the liquid.[42] However, most instructions outline the traditional method of kneading the crushed stone under water with a paste of wax, resin and oil so that blue particles float into the water while the unwanted minerals are retained by the paste. Instructions for that method are noted by de Mayerne, and Norgate gives extremely full details of the process. After describing the selection and preparation of the stone (including the removal of grey impurities by means of pincers, a point which others do not emphasise), Norgate describes how the stone must be ground in water, and the powder allowed to dry. Then he describes the preparation of the paste into which the powder is worked and afterwards allowed to stand for a day or two, and finally gives instructions for the lengthy kneading operation by which means various degrees of blue pigment are produced:[43]

> Then take a ffayre Earthen pann, or a Cleane smoth basson allmost full of water, soe warme as yow may well indure your hand in it.
> Then take this Mase or lumpe [of paste and powder] and kneade it and worke it betweene your handes soe longe that yow may see it sweate out drops of cleare water of a blew Colloure, and the longer it is before the drops come forth, the better. When your water is well blewed, sett it away, and take an other Basson or panne of cleane water warme as before, and worke the same lumpe as before, and then take an other pann in like sorte, and soe a fourth and a ffifte, till yow see that noe more Blewe water drops will sweate out.
> Lett this water stand and settle 24-howers, and then power it off and lett the ground[s] drye, and when it is thoroughly dryed, wipe it out of the Pann with a ffether uppon a paper and soe putt it upp.
> Note that the first Blew that sweateth out is the best, the next a second sorte, and soe a 3 or 4. Yow may put the sleighter of those sorts into the like newe pastill agayne, and worke it ouer as before, and soe yow shall have it fayrer than it was, but lesse in quantaty.
> Note allsoe that the Pastill cann never serve but once, and serveth afterwardes for noe use but to make linkes or Torches. Yow may gett some of the broken peeces of Lapiz Lazuli of the Merchantes of Alleppo, the deepest is the best.

It was common practice to continue the process until nothing but a pale blue-grey was obtained. Norgate does not indicate what quantity can be expected, but another recipe contains the information that from one pound of good quality stone four to five ounces of the best, three ounces of medium and about three ounces of the weakest ultramarine should be obtained.[44] In later years the weakest variety was called *ultramarine ashes* owing to its grey appearance, but the term was not commonly used in the seventeenth century.

At all times the demand for natural ultramarine greatly exceeded supply, and, for that reason, it was the most costly of all pigments. The price of eleven pounds ten shillings per ounce for the best and seven pounds ten shillings per ounce for the poorest quoted by Hilliard, *circa* 1600, may be compared with ten guineas per ounce for the best and two pounds per ounce for ultramarine ashes quoted in *The Artist's Repository* almost two centuries later. Such prices tend to assume too much importance, as, for the painter in water colours anyway, it was not always necessary to buy a large amount and the 1701 edition of John Smith's book includes a small quantity of pigment at a price of only two shillings. Supply of small quantities at a reasonable price enabled ultramarine to be used by miniature painters, but the pigment was generally considered too costly for use in washing prints and maps. Eighteenth-century writers on oil and water-colour painting refer to ultramarine, but documentary evidence suggests that it was used less then than in the seventeenth century. A revival in the use of the pigment in the nineteenth century was probably the result of a new source of supply combined with the initiative of a London colourman, both of which are mentioned by Ibbetson in discussing ultramarine:[45]

> It is made of the beautiful marble called Lapis Lazuli, which ranks among precious stones, and the preparation is difficult; but is now to be had in perfection of Middleton, in St. Martin's Lane. Quarries of this precious substance have been discovered in the Russian territories, bordering on Persia. A Russian agent was some years ago in London, who offered to procure it in quantities, five hundred weight if wanted, but could get no orders; until Middleton, who has great spirit, and knowledge of colours, has undertaken the manufactory of it, and has it of all degrees of value or depth. I have used his middling sort which is excellent for landscape.

Amongst several examples of natural ultramarine in one of Field's manuscripts, the 'Practical Journal 1809', is a water-colour wash made up from pigment obtained from Middleton at six guineas per ounce; the pigment proved to be permanent in both oil and water colour in Field's fading tests.[46] In *Chromatography,* Field mentions lapis lazuli from Siberia, so it is likely that stone from the mine near Lake Baikal in Russia contributed to the greater availability of ultramarine in the nineteenth century.

As 'the diamond of all colours', ultramarine received enthusiastic

Fig. 16. Oil-colour box by the colourman Thomas Brown, High Holborn. The bladders used as containers were pierced with a tack in order to squeeze out the colour. This box pre-dates the introduction of collapsible metal tubes that were patented by John Rand and sold by Thomas Brown in 1841. (Reproduced by kind permission of Messrs Winsor and Newton Limited.)

praise and an almost legendary reputation. A number of eighteenth-century writers were more interested in recounting an anecdote concerning a large quantity given away by Charles II than in discussing the actual qualities of the colour, suggesting therefore that they were possibly unfamiliar with the colour because of its rarity. Bardwell comments on its translucency (a characteristic which is shared by blues in general), saying, 'It is a tender retiring Colour, and never glares; and is a beautiful glazing Colour: It is used with Poppy oil.' Some writers point out that ultramarine oil colour tends to fatten and become unusable if stored for long in bladders, and Field states that ultramarine does not wash well in water colour. Nevertheless, most writers found nothing but praise for the colour, and its hue and good degree of permanence to light set a standard by which more modern blues were judged, for a desirable recommendation for a new blue was its ability to replace ultramarine.

Azurite: $2CuCO_3 \cdot Cu(OH)_2$

Azurite is the modern name for a blue mineral, a naturally occurring basic copper carbonate which is found in close association with the green mineral, malachite. In the seventeenth century azurite was commonly called *lapis Armenius,* as that was the name used in classical antiquity. The pigment prepared from the mineral was named blue bice, although occasional references to the pigment as mountain blue are to be found. Seventeenth-century names connected with the mineral and pigment are mentioned by Waller:[47]

> I take the Lapis Armenius to be the blew Bice sold in the Shops, for it is light and friable; formerly brought out of Armenia, now from the Silver Mines of Germany, called Melochites, in high Dutch Bergblaw.

All the names were understood and properly used by writers on painting in the seventeenth century, and it was not until following centuries when the pigment was little used that there was confusion over the names. Lapis Armenius was misinterpreted as lapis lazuli or blue ochre, and bice was commonly applied to other blue pigments. The English name bice is somewhat remarkable in that it has been applied to different pigments at different times, and, as the seventeenth-century name for the blue prepared from azurite, it is unlike other European names for that colour at the same period; it bears no relationship to mountain blue, Teutonic *bergblau,* nor to blue ashes, French *cendres bleues* and Spanish *cenizas de azul.* Amongst them bice stands alone as a name of obscure derivation.[48] There is, however, one certain feature in that, whatever the name meant at an earlier or later date, in the seventeenth century the name indicated blue from azurite and no other pigment.

Seventeenth-century sources contain far more references to the place of origin of azurite than to that of ultramarine; sources mentioned include Hungary and Germany. Haydocke doubtless meant azurite when he referred to ultramarine of Hungary, but such confusion between bice and ultramarine is rare. Supplies from Hungary were interrupted as a result of Turkish invasions in the sixteenth century, and they never really recovered.[49] Consequently, Germany is mentioned most frequently in English books. Another seventeenth-century source which has previously escaped attention is Central America. B.M. MS. Sloane 6284, f. 109v, includes both *bise* and *Spanish bise* in a colour list, and four references to bice from the Indies are made by de Mayerne.[50] Previous writers have interpreted his reference as meaning azurite from the East Indies, but when it is described in one place as being imported to Spain and in another as found 'aux Indes occidentales', it is absolutely certain that it was of American origin, probably from New Mexico where azurite is known to occur.

In its raw state the mineral may be a deep blue and appear somewhat similar to lapis lazuli, but the pigment is extracted in a completely different way. De Mayerne proved that his American blue was actually bice, for attempts to extract the pigment by the method used for ultramarine were unavailing, whereas another, successful attempt followed an accepted process for obtaining bice from azurite.[51] The usual method was to crush the mineral and then wash it. Sometimes it would be washed with vinegar to remove any green impurity, and then extensive washing with water followed. Traditionally, honey, fish glue or gum was added to the water to induce the blue pigment particles to separate from the dross quickly and sink to the bottom of the vessel. The following instructions, which are quoted from Peacham, were copied verbatim by Bate some years later.[52]

> Blew Bice. Take fine Bice and grinde it vpon a cleane stone, first with cleane water as smal as you can, then put it into an horne and wash it on this manner: put vnto it as much faire water as wil fill vp your horne, and stirre it well, then let it stand the space of an houre, and all the bice shall fall to the bottome, and the corruption will fleet above the water, then powre away the corrupt water, and put in more cleane water, and so vse it foure or fiue times, at the last powre awaie all the water, and put in cleane water of Gumme Arabick not too stiffe, but somewhat weak, that the bice may fall to the bottome, then powre away the Gumme water cleane from the bice; and put to another cleane water and so wash it vp, and if you would haue it rise of the same colour it is of, when it is drie, temper it with a weake gum water, which also will cause it to rise and swell in the drying, if a most perfect blew, and of the same colour it is being wet, temper it with a stiffe water of gumme lake, if you would haue it light, grind it with a little Ceruse, or the muting of an hawke that is white, if you will haue it a most deepe blew, put thereto the water of litmose.

Bice was frequently merely mixed with a binding medium without grinding, as, in common with some other pigments, it tended to become very pale if finely ground. As Norgate explains, 'if yow thinke to make them fine by grinding they instantly loose theire bewty, becoming starved and dead.'[53] In contrast, however, John Smith states that the pigment needs good grinding.

Blue bice sometimes had a slightly green cast, which made it unsuitable for certain purposes. Miniature painters would not use it for flesh tones: 'ffor a blewe Shaddow Indico and white (for blewe Bice is never used in a fface).'[54] Norgate was similarly opposed to its use in painting skies: 'work your Blewish Skyes and Cloudes with Smalt, not with Bise, for it is too Greene and Blewe, and noething soe proper for the purpose.'[55] Nevertheless, it was useful and was often recommended as the best blue to use in place of ultramarine in water-colour painting. A green tendency could also be troublesome in oil painting. The pigment was generally tempered on the palette with nut oil, because it yellows less than linseed. Use of spike oil in conjunction with the oil colour was in de Mayerne's time reputed to prevent the pigment sinking in the oil in a paint layer and so counteract a visual appearance of green which might develop if a blue pigment was covered by a slightly yellow film. Despite these difficulties, bice was a useful colour and often recommended as the best blue to use in place of ultramarine. It was a moderate price when compared with that pigment, but at a price of six shillings and eightpence per pound, quoted in V. & A. MS. 86.EE.69, it was more expensive than many other colours. Judging by the frequency with which it was mentioned during the first three-quarters of the seventeenth century it was used extensively, but during the later part it was considered too good for some purposes.

It is impossible to give a definite opinion as to when the colour name bice ceased to mean the colour prepared from azurite. It is fairly clear that the pigment fell into disuse towards the end of the seventeenth century (it is not mentioned by Marshall Smith), but use of the name continued. In the eighteenth century bice was understood to mean a very pale blue, and several references (including one by Dossie) indicate that finely ground smalt was sold under that name. In the nineteenth century Field knew that bice was an obsolete pigment for which both smalt and verditer had been substituted, but he did not connect it with azurite. Ure ignored its earlier history completely, and his statement that bice is a pigment prepared from smalt has been taken as authoritative in the twentieth century, with the unfortunate result that the seventeenth-century colour name is open to misinterpretation.[56] The fact that nineteenth-century writers did not connect bice with azurite does not mean to say that azurite was then obsolete. It fell into disuse in the eighteenth century but was revived in the nineteenth, for, even though it may not have been used extensively then, it was available and was listed in colourmen's catalogues as azurite.

Manufactured Copper Blues

Blue verditer, the manufactured equivalent of azurite, was a relatively new colour in the seventeenth century, although other manufactured blue pigments containing copper were known and used long before its introduction. Seventeenth-century artists were not really concerned with any copper blues other than azurite and verditer [$2CuCO_3{\cdot}Cu(OH)_2$], but some comments on traditional, artificial azures must be included here, as numerous recipes for making them appear in sixteenth-century sources and seventeenth-century copies.[57] Many give verdigris as an ingredient, whereas others suggest the use of brass, silver or quicksilver. Verdigris is a basic copper acetate, and brass a copper alloy. Silver may have contained copper as an impurity, for there is reason to suppose that silver refining was not always absolutely successful in practice even though it was well understood in theory. Traditional copper blue recipes sometimes specify that the ingredients should be prepared in a brass vessel, in which case the container would have provided the colouring matter. Many of the workable recipes depend on the use of sal ammoniac (ammonium chloride) to form a blue compound with the copper, and many suggest the use of some form of lime, such as powdered eggshell, to give body to the colour. Thus, the blue pigment produced by most of the traditional methods would have been a cuprammonium compound containing lime.[58] Some recipes are impracticable; quicksilver and sulphur are recommended as ingredients for making azure, but the instructions are the same as those for making vermilion. English documents offer no clue as to why a red pigment should have been confused with blue in the Middle Ages.

Written instructions for making copper blues other than verditer are archaic even by early seventeenth-century standards, and it is clear that they do not represent seventeenth-century practice in colour manufacture. The pigment is generally named azure or byce, even though at the beginning of the seventeenth century blue was the current name for the hue and azure was restricted to heraldry or poetic language. Similarly, bice was the name applied exclusively to azurite by Peacham, Norgate and de Mayerne, all of whom are reliable authorities for the first half of the seventeenth century. It is likely that the bice mentioned by Hilliard was also azurite. Of the four, only de Mayerne mentions manufactured copper blue other than blue verditer, and it is evident that he copied a number of instructions from other sources, because these instructions are in English or Italian, languages that he did not normally use. Without exception, the pigments produced by following all the relevant English recipes are called azure; significantly, they are called neither bice nor blue. The Italian recipes come from a book printed in the first half of the sixteenth century.[59] The appearance of such recipes in de Mayerne's manuscript and in seventeenth-century copies of earlier sources is merely evidence of antiquarian interest, for the absence of cuprammonium blues from English sources which were unquestionably composed during the first half of the seventeenth century and convincing evidence that bice was then the name for azurite point to the conclusion that such artificial azures were obsolete by 1600. The traditional recipes were originally intended as instructions for small quantities of pigment to be made by individuals. Documentary evidence concerning blue verditer, the seventeenth-century manufactured copper blue, shows that it differed from the old pigments in chemical composition and in its method of manufacture in large quantities by an industrial process.

It is partly the greater frequency of mention of cuprammonium lime blue pigments as opposed to blue verditer in sources on artists' pigments that suggests that verditer was fairly new in the late sixteenth century. There were two varieties of verditer, blue and green, and the name, that was probably derived from the old French *verd de terre,* indicates that the green variety was produced first. Hilliard and Peacham mention green verditer but not blue, whereas Norgate mentions both. Although the name was derived from French, it was entirely English in application, as the French used *cendres bleues* for both native and manufactured blue copper carbonates. Confusion between the two was sometimes avoided by use of the name *cendres bleues d'Angleterre* for blue verditer because it was regarded as an English speciality, but there was a certain amount of confusion in England in the eighteenth century when, following a fashion for French customs and language, the name Sanders blue was introduced and used as if for a separate colour. The corrupt form of *cendres* met with considerable criticism from Dossie when he found that the pigment sold by that name was nothing other than verditer.

The circumstances of the discovery of verditer are unknown, and, although the similarity between it and bice was recognised in the seventeenth century, it is almost certain that verditer owed its origin to chance and not to the analysis of a traditional pigment followed by the synthesis of a new one, as was the case with a number of other modern colours. De Mayerne's information that the pigment was discovered accidentally probably contains much truth. He was told that someone had taken the liquid (actually copper nitrate) which was produced during the process of parting silver from copper and had by chance thrown it on to chalk or ceruse which had at once turned green.[60] Later in the seventeenth century Boyle and Merret separately investigated the manufacture of verditer, which was by then known as a by-product of silver refining, and both commented on unpredictable results. According to Boyle, the refiners did not understand why the process was sometimes unsuccessful, but he does not explain the nature of the difficulty and leaves the impression that the manufacturers sometimes produced no pigment for months at a time.[61] Merret's comments show that the difficulty lay in obtaining a blue pigment because more often than not the product turned out to be green:[62]

> Tis a strange and great mystery to see how small and undiscernable a nicety . . . makes the one and the other colour, as is daily discovered by the refiners in making their Verditers, who sometimes with the same materials and quantities of them for their Aqua-fortis, and with the same Copper-Plates, and Whiting make a very fair Blew Verditer, otherwise a fairer or more dirty-green. Whereof they can assign no reason, nor can they hit on a certain rule to make constantly their Verditer of a fair Blew, to their great disprofit, the Blew being of manifold greater value than the Green.

Modern research has made it clear that refiners' blue verditer may be made successfully at a lower temperature and with more frequent and vigorous stirring than its green counterpart and that, even in the cool climate of the British Isles, a spell of relatively warm weather would account for interruption in the production of blue verditer.[63]

Eighteenth-century English sources add very little to the information given by Boyle and Merret, whose accounts probably form the basis of those written later. The pigment was made on a considerable scale, 100 pounds of whiting being used at a time in a large tub.[64] As the liquid which remained after completion of the colour-making was reclaimed and used again in the parting process, the manufacture of verditer cost the refiners nothing apart from the labour and the negligible price of the whiting. Throughout the seventeenth and most of the eighteenth century, manufacture was attributed solely to the refiners. It may have endured for a long time as a profitable side-line in the metal industry, and there is evidence in the records of the Sheffield Smelting Company to show that John Read's business took up paint manufacture in 1781. Green verditer was sold at a price varying from ninepence to one shilling and sixpence per pound, and blue verditer was more expensive at four shillings per pound.[65] The retail price of blue verditer at about the same period was sixpence to eightpence per ounce.

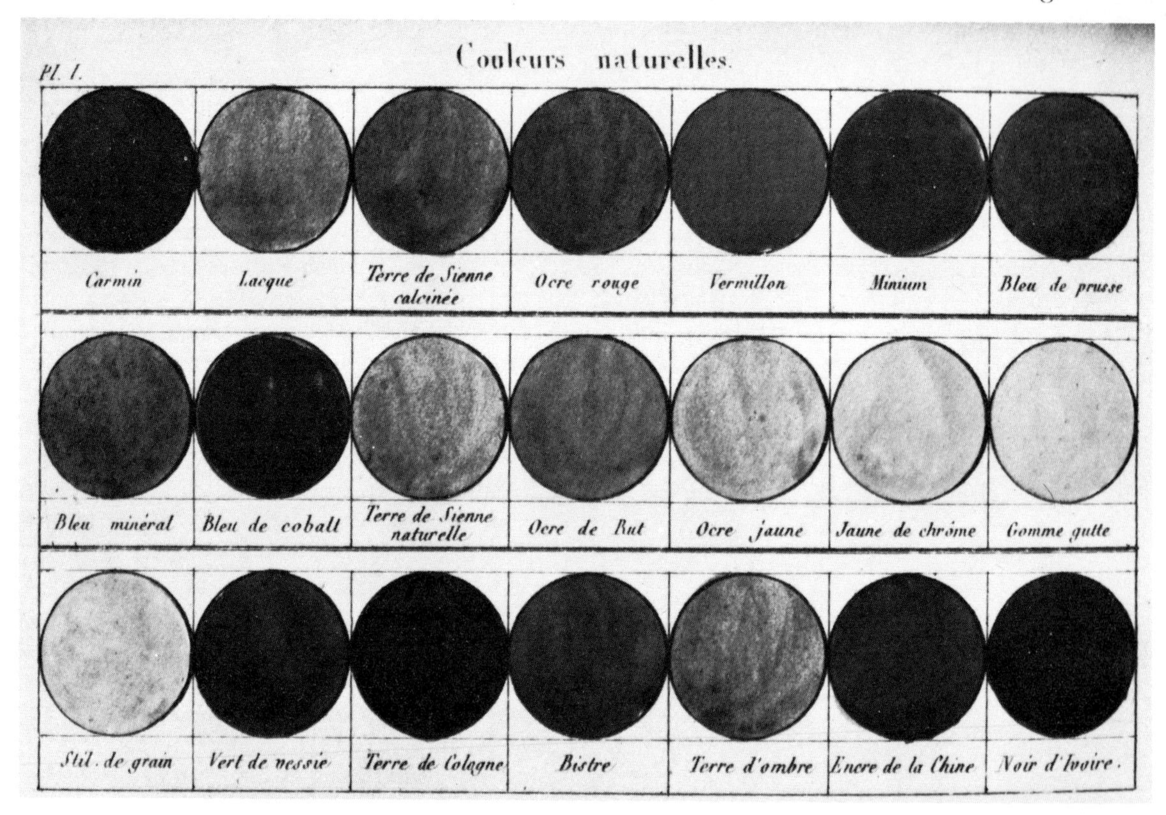

Fig. 17. Early nineteenth-century French colour samples, including mineral blue (manufactured copper blue), cobalt blue and chrome yellow (*Le maître de miniature, de gouache et d'aquarelle*, 1820). (Reproduced by courtesy of the Science Reference Library.)

The manufacture of blue verditer was not properly understood in France until the late eighteenth century, when Pelletier analysed the pigment. According to Chaptal, a French industrial chemist, Pelletier's method of making the colour consisted of mixing copper nitrate and lime, washing the green precipitate and mixing it, while wet, with quicklime so that the resulting pigment was blue. Chaptal put forward various improvements on that method, suggesting the use of limewater or baryta water instead of ground lime. He obtained a constant result and claimed that the elimination of a carbonate from the composition of the pigment ensured that it was not gritty and that quick drying in the dark and the removal of all acid helped the blue to retain its colour. Chaptal also pointed out that the triple salt formed when ammonium chloride was dissolved in baryta water gave a fine blue.[66]

In nineteenth-century English sources, Field and Ure both state that verditer contains lime and Field describes the pigment not as copper carbonate but as copper oxide. It is clear that various copper blues that were different from refiners' verditer were manufactured in the nineteenth century. It has been suggested that changes in the manufacture of copper blue pigments around this time resulted from a change in the refiners' parting process during the eighteenth century when sulphuric acid was used in place of nitric acid.[67] It seems that the various copper blues that were manufactured were not only different from traditional refiners' verditer but were also inferior to it.

Blue verditer itself was never widely employed in artistic painting, owing to two defects: its green cast tended to increase on exposure and its texture was sometimes coarse. Norgate lists blue verditer not with other blue colours but amongst greens, obviously intending it to be used only in mixture with yellow. Gyles excludes both blue and green verditer from water-colour painting on the grounds that they are too gross and coarse, but they are admitted to his list of oil colours. In spite of its disadvantages, blue verditer is mentioned fairly often in sources dating from the second half of the seventeenth century, references to it increasing as those to blue bice become less frequent. The texture of blue verditer may have varied but it was not always coarse as is borne out by the sample included in Field's 'Practical Journal 1809', f. 316, where he states that the blue verditer purchased by him from Smith & Co. for one shilling per ounce was 'in fine soft powder, works very even in water'. The cheapness of blue verditer was an important point in its favour, and, for that reason, it was recommended as a colour for washing prints. Another result of its small cost and bulk production was its use as a decorators' paint, and it is likely that it was far more important in decorating than in artistic painting. Professional painters would not use it if anything better was available. It was never popular, although it may have fulfilled a certain demand at the beginning of the eighteenth century when azurite was scarce and Prussian blue was not yet available.

Smalt:[*] K, Co(Al), silicate (glass)

Of the many colours which can be made from cobalt, smalt was the only pigment to be manufactured before the isolation of the element in the eighteenth century. Cobalt ore was heated to form an oxide which was mixed with silica and, under the name saffer or saffre, was then sold to glass-makers who fused it with potash to form the blue glass known as smalt.

Cobalt occurs in the form of various ores in many parts of the world, and that used for colouring glass in early times was probably mined in the Near East. The history of cobalt mining in western Europe is obscure, although the ore known as skutterudite, and sometimes as smaltite, which is found in Saxony in association with silver, arsenic and bismuth, was for centuries the main source of European cobalt. Mention is often made of the fact that the name *cobalt* was applied to metallic ore and the goblins reputed to inhabit the mines, an enchanting association until the reason for it is known: the goblins hindered the miners' work, and the name was transferred to a dangerous ore which was likewise a hindrance. In the sixteenth century miners working in close proximity to cobalt arsenide were advised to wear leather masks and protective gloves and boots to prevent their flesh being eaten away by arsenic, details of which are provided by Agricola, a physician who worked amongst the miners.[68]

Traditionally, cobalt ore was first worked in the mines at Schneeberg *circa* 1470, which was a little before the time of Agricola. Fifty years later Weidenhammer, a Frenchman, settled there and began making saffer, and Schürrer was manufacturing smalt at Neudeck *circa* 1540.[69] These facts are probably correct, but lack of documentary evidence may have led to a belief that both cobalt mining and saffer manufacture were introduced later than was actually the case. Blue glass was of course mentioned long before the sixteenth century, and there is always the possibility that cobalt could have been brought from the Near East during the Middle Ages or that knowledge of how to use it came from the same source. However, there is a modern view that cobalt glass was made in

Fig. 18. Cobalt smelting in Saxony during the seventeenth century: A is the beginning of the tunnel through which the fumes passed and arsenic settled; B is the reverberatory furnace in which the cobalt ore was placed (J. Kunckel, *Ars vitraria experimentalis*, 1689). (Reproduced by courtesy of the British Library.)

Europe at least as early as the fourteenth century and that it may have been an Italian invention and not German as previously supposed. The suggestion has been put forward on the basis of a description by Antonio of Pisa, written during the fourteenth century, concerning glass made from *chafarone* brought from Germany.[70]

Whatever the date of its discovery, saffer was certainly known throughout Europe in the sixteenth century, but its composition was a matter for conjecture by a number of writers until the second half of the seventeenth century when the method of manufacture was finally explained in a written and illustrated description by Kunckel, superintendent of glassworks in Saxony. According to his account, cobalt ore was placed in a reverberatory furnace and the fire was built up so that flames met over the top (*Figure 18*). Arsenic contained in the ore was driven off in the form of white smoke which was channelled into a long wooden tunnel where a large proportion of it settled so that it could be collected later. When no further arsenical fumes appeared, the roasted cobalt was removed and ground and the whole process was repeated. Later the cobalt oxide was sieved, mixed with pulverised stones, moistened and packed in barrels where it set stone hard so that it had to be broken up by sledge-hammers before it was used. Kunckel's account is written in German, but a seventeenth-century English translation exists in manuscript form in the Royal Society Classified Papers.[71]

Kunckel's remark that smalt can be made by adding sand and potash to saffer and heating the mixture so that it vitrifies is amplified in an eighteenth-century French edition of his work. Equal parts of saffer, silica and potash were mixed and placed in pots in a glass furnace where they were left for twelve hours, the mixture being stirred at intervals after the first six hours. The vitrified mass was then removed by means of a ladle and plunged into water to make it friable so that it could be sieved, ground and finally washed. It was packed in barrels marked according to colour grade, the best being a deep violet-blue and the cheapest a pale powder-blue.[72]

Smalt was manufactured in the Netherlands in the sixteenth century, and the Dutch or Flemish pigment acquired a reputation for excellent quality. Smalt-making began in England at the very beginning of the seventeenth century and was practised for a time by a British manufacturer and a Dutch immigrant. Details of Dutch opposition to the British industry and the unfortunate rivalry between the manufacturers in England are contained in Appendix Three. The native industry failed and had to be recommenced at a later date, its failure being the result of the monopoly system, not of a lack of demand, as there was a ready market not only amongst painters and decorators but also amongst potters, paper-makers, paper-marblers and laundresses, the last of whom used a very pale variety of smalt as a washing blue.

English painters were familiar with smalt some time before it was manufactured at home. It was with reference to smalt that the author of *Limming,* 1573, issued a warning that the apothecaries were liable to mix sand or chalk with the pigment 'to multiplie it to theire profit'. Artists used high grade smalt of a deep blue with a purple cast, which made it useful for obtaining mixed purples in conjunction with red lake, and when used alone many painters, including Hilliard and Gyles, considered it suitable as a substitute for ultramarine. A great defect of the pigment was its tendency to become markedly paler if ground at all finely. Norgate lists smalt as one of the pigments which should be washed but never ground, and de Mayerne notes that it must be tempered very lightly on the palette because it fades if it is worked with a knife.[73] Documentary evidence suggests that smalt was used quite extensively in oil painting in the seventeenth century but less in the eighteenth. In oil, pigment particles tended to sink in a paint layer, so in decorative work, such as that described by John Smith, the pigment was often scattered over the tacky surface of oil paint. Good effects could be achieved by that method, because the pigment, being glass, tended to catch the light and sparkle slightly, but the technique was not always acceptable in easel paintings. Some oil painters (Marshall Smith, for example) recommended that smalt was best tempered with oil and then used for glazing purposes. Smalt was obviously used by seventeenth-century painters in water colours, but it was not particularly easy to work with as it was impracticable to grind the colour finely and it did not flow well. 'Somewhat thick and clodding' was the verdict of Goeree.[74]

Smalt survived throughout the eighteenth century as a pigment of medium price which was used in decorating and in water-colour painting of no great importance. It was still supplied in the nineteenth century; in fact, Field tested two pigments that were almost certainly smalt, one of which, from Brandram & Co., was sold as royal blue, and another obtained from Newman under the name French smalts that Field also listed as Dumont's blue. He found them deficient, as in oil they discoloured badly in damp, impure air.[75] However, by the time Field carried out his fading tests, scientific research into the nature of cobalt had replaced smalt with a blue of equally good colour and much better properties.

Cobalt Blue: $CoO.Al_2O_3$

Conclusive proof that the blue colour of smalt was due to the metal, cobalt, and not to other substances generally associated with the ore was not obtained until the first half of the eighteenth century when Brandt, a Swedish chemist, isolated the element and found that it was the source of blue colour. Research on cobalt compounds followed later in the century, and both Gahn and Wenzel separately discovered the blue compound, cobalt aluminate, which was later recommended for practical purposes by Thenard. He was undoubtedly aware of previous research, as his book *Traité de chimie* contains an impressive list of references to earlier scientific papers on the subject.[76] The initiative for his own investigation came from Chaptal, who, in his capacity as minister in the French government, appointed Thenard and Mérimée to investigate the possible improvement of artists' colours. According to both

Chaptal and Mérimée, the discovery of cobalt blue took place in 1802 although Thenard's report on his experiments was not published until the end of 1803.[77]

The fine blue of Sèvres porcelain gave him the idea of experimenting with cobalt, and various trials showed that when cobalt arsenate or cobalt phosphate was combined with alumina by roasting at a high temperature, a good blue was obtained. When Thenard made his report, trial samples of the pigment had been tested satisfactorily in an oil and gum medium and the colours were unchanged after exposure tests which had then lasted two months.[78] Cobalt blue went into production without delay in France, as, in 1807, Chaptal commented that it had already become a branch of commerce. The French pigment is generally regarded as having been cobalt phosphate containing some alumina, whereas the pigment made by Leithner at Vienna was composed of cobalt arsenate. Some modern authorities appear to use the names Thenard's blue and Leithner's blue to distinguish cobalt blues of slightly different composition, although nineteenth-century writers generally referred to any variety by the simple name, cobalt blue.[79]

It is difficult to determine the date of the introduction of cobalt blue in England, but it is unlikely to have been much later than in France, as French scientific publications were frequently translated into English and there was obviously an interchange of ideas even while both countries were at war. The necessary raw materials for making the pigment were available in England, and there is reference to the use of cobalt mined in England in the manufacture of a much improved blue for porcelain in 1815.[80] It is listed as cobaltic blue in the earliest example of the colour in Field's 'Practical Journal 1809' in an entry that probably dates from soon after 1815. He mentions that his pigment samples were obtained from Bollmann (who is known as a colour-maker in England at this time), that they were manufactured according to the method of Berzelius and that Newman was selling the pigment at one guinea per ounce.[81]

Cobalt blue is not included in the list of colours recommended by Varley for use in landscape painting which was printed with his permission in Hamilton's book of 1812, but it heads *J. Varley's List of Colours* of 1816, so it seems that the artist must have adopted the colour in the interval. Cobalt blue is mentioned with approval in that painter's list and in several other books of the early nineteenth century. Varley recommends it as a good substitute for ultramarine and a better colour for skies where even tints are required. He remarks on its brilliance, a characteristic which is also mentioned by Dagley, who says that it tends to overpower other colours but that its permanence outweighs any possible disadvantage. Field was not so certain of its durability, saying that impure air made it turn first green then black. The pigment is generally regarded as durable, but Field's comment on the pigment as prepared in the early nineteenth century is certainly borne out by the condition of the colour sample in a copy of *Le maître de miniature,* 1820. In spite of a few criticisms and early comparisions between ultramarine and cobalt blue, the new colour was adopted by artists not so much as a good substitute but as an excellent colour in its own right.

French Ultramarine: $Na_{8-10}Al_6Si_6O_{24}S_{2-4}$

A true substitute for ultramarine from lapis lazuli was found at last after analytical experiments had shown that it was composed of soda, silica, alumina and sulphur. That knowledge, combined with the observation of a blue substance corresponding to ultramarine in a furnace which had been used for soda manufacture, concentrated attention on the possibility of making a synthetic ultramarine, and general interest in the early nineteenth century was reflected in prizes offered as an incentive to research. The *Transactions* of the Royal Society of Arts for 1817 contains an offer of a premium of thirty guineas or a gold medal to anyone who could submit evidence of having prepared an artificial ultramarine which was equal to the best prepared from lapis lazuli but cheaper in price. There were no claimants, but a much more generous prize of 6000 francs (then about £500) was offered in 1824 by the Société d'Encouragement pour l'Industrie Nationale in France, and in 1828 it was awarded to Guimet, a French manufacturer who did not reveal his method. The award prompted Gmelin, a German, to publish a process which he had discovered independently the previous year. He explained how he obtained silica and alumina, prepared a solution of silica with caustic soda and then heated the solution together with alumina so that a dry powder was obtained. A mixture of two parts sulphur to one of sodium carbonate was heated in a crucible until it was red, and then the powder was added to it gradually and the whole was kept at a red heat for one hour, after which it was allowed to cool. The remaining substance was ultramarine which required to be washed in order to remove an excess of sodium sulphate. Sometimes an excess of sulphur had to be removed by applying gentle heat. The publication of Gmelin's method in a French journal was accompanied by a letter from Guimet claiming that he first manufactured ultramarine in 1826 and that many eminent painters, including Ingres, had used the pigment in the following year. Ingres employed the new blue in a ceiling painting at the Musée de Charles X and reported that it left nothing to be desired.[82] In 1830 Guimet published further comments, giving the address of his distributor in Paris and pointing out that manufactured ultramarine was available in a superfine quality for artists and a second quality for other purposes, such as the production of blue tinted papers by Montgolfier, the manufacturer of the famous Canson marque. Guimet emphasised the fact that even his best quality pigment was cheaper than cobalt blue, which he thought it might replace.[83]

It is clear that ultramarine was immediately produced on a commercial scale in both France and Germany and it was imported to England where it was generally known as French blue or French ultramarine, although the name German ultramarine sometimes appears in early nineteenth-century sources. The long-awaited substitute for natural ultramarine was not acclaimed by the British in quite the way one might expect. Field, who was obviously familiar with different varieties of artificial ultramarine, stated that none possessed the merits of natural ultramarine. In his opinion their value had not been fully established by 1835. Fielding put forward a similar view in *Ackermann's Manual* of 1844: artificial ultramarines were useful but less pure and less permanent than the pigment obtained from lapis lazuli.

Miscellaneous Inorganic Blues

Haarlem ultramarine is mentioned in the later version of Norgate's treatise where it is recommended as useful in landscape painting but its composition is not mentioned. Eastlake refers to Hoogstraten's mention of Haarlem ashes, suggesting that it was probably a copper-based manufactured blue. He was right in stating that, whereas ashes indicated ultramarine ash blue in the nineteenth century, it usually indicated blue verditer or similar copper-based blue at an earlier period. Nevertheless, in recent research concerning Richard Symonds, Beal has pointed out that that seventeenth-century writer describes Haarlem ultramarine as 'a blew clay earth that is washt . . . & tis not any way produc'd from Lapis Lazuli', suggesting, therefore, that it was of natural origin.[84]

It seems possible that some unusual inorganic blue pigments were in occasional use and that they were little mentioned until scientific papers were printed in the eighteenth century. An example is a blue-black or blue earthy substance, found in Germany, that is named ilsemannite after the scientist who published a study of molybdenite in 1787. It is interesting to note that the colour-manufacturing firm of Berger listed over one hundredweight of molybdic (of unspecified colour but listed with blue verditer and various green pigments) in their stock in 1805.[85] The blue pigment is referred to in some sources as molybdenum blue or blue carmine (on account of its reputedly good flow when used as a water colour). It is mentioned under the latter name by Field, but he obviously had no first-hand knowledge of it or its composition. It is probably a complex oxide of molybdenum, generally described as colloidal.

More frequent references are made to blue ochre in late eighteenth-century and early nineteenth-century sources. Octahydrated ferrous phosphate [$Fe_3''\ (PO_4)_2.8H_8O$] or vivianite, as it is now known, occurs in many places, including Cornwall where it occurs as a secondary mineral in association with tin. It was called blue ochre or native Prussian blue, because it was known to contain iron.[86] It seems possible that it was used experimentally in English painting, as Field was able to test a sample of the pigment imported from America that he found was unaffected by sunlight or impure air. In *Chromatography* he states that it is difficult to obtain but that it works well in water colour and oil.[87]

5

Organic Blues

Organic colours are compounds of carbon with oxygen, hydrogen, nitrogen, sulphur and other elements; they were in the past prepared from animal or vegetable substances. Traditional organic blues, that is, those in use before the eighteenth century, were made by extracting dye from various plants. Some of those dyes turn blue or violet after special treatment, but in their natural state they are red. For that reason, colours such as turnsole and orchil are listed as reds in some books, as in *Colour Index,* 1956, the standard work on traditional and modern dyes which has been used as an authority for the botanical names given below.

Turnsole

The dye expressed from the berries and tops of *Croton tinctorium* was traditionally used to stain pieces of cloth which were then treated with an alkali, generally by exposing them to vapours of ammonia, so that they turned violet or blue. Turnsole-dyed cloths were certainly known in England in the seventeenth century but, according to the herbals written by Gerard and Parkinson, the plant, then called *Heliotrope tricoccum,* grew only in France, Spain and Italy; there was not enough sunshine in England. Gerard, once Master of the Barber-Surgeons, describes the cloths as follows:[88]

> With the small Tornsole they in France do die linnen rags and clouts into a perfect purple colour, wherewith cookes and confectioners doe colour iellies, wines, meates, and sundry confectures: which clouts in shops be called Tornsole, after the name of the herbe.

He generally mentions if a dye is used in painting, but he does not refer to turnsole in that connection.

The dye was certainly not used in oil painting and it was probably little used in water colours by the seventeenth century. The colour is mentioned in some manuscripts copied from earlier sources, and turnsole tempered with glair is mentioned in *Limming,* 1573. A water-colour sample, that has reverted to a lilac colour, is included in de Mayerne's

Fig. 20. Turnsole plant, as illustrated in Parkinson, *Theatrum Botanicum*, 1640. (Reproduced by courtesy of the Royal Horticultural Society.)

4. *Heliotropium tricoccum.*
The colouring or dying Turnesole.

manuscript, B.M. MS. Sloane 2052 (*Plate 8*). Norgate deliberately omitted turnsole from his list of recommended colours, naming it with others 'which are extraccions from fflowers, Jvoyce of herbes or Rootes, which by reason of theire cheapenis and Communitie are esteemed fitter for those which wash prints, Cartes, Mapps then for Limning'.[89] The point was that the organic colours named were undesirably transparent and insufficiently light-fast. Some of the later writers who copied Norgate mention turnsole for washing, and both Peacham and the author of *The Excellency of the Pen and Pencil* describe how turnsole rags may be boiled with vinegar and then tempered with gum. After such treatment with an acid, the dye would have reverted to red, which explains Peacham's remark that the colour is good for shading flesh tints and yellow.

A final reference to the colour as sunflower blue appears in de Massoul's book written at the end of the eighteenth century, and there is evidence that small quantities of turnsole were imported from Italy during the same century, but the dye was probably used for purposes other than painting, for it is clear that turnsole was no longer in regular use as an artists' colour at that time.

Orchil, Cork and Litmus

Various lichens of the *Roccella* group supply a dye which has in the past been used in textile dyeing and to a lesser extent in painting. The lichens grow in many places, some of them in rocky parts of the Welsh and Scottish coasts, where they were collected during the seventeenth century and probably earlier. When Ray, a Fellow of the Royal Society, visited Malta in the second half of the century he remarked on a moss called *vercella* which was found there and used as a dye and he compared it with the British variety: 'This kind of moss, called in Wales *Kenkerig,* and in England *Cork* or *Arcel,* is gathered and used for the same purpose in Wales and the North of England.[90] Possibly the native product did not grow abundantly enough because by the eighteenth century large quantities of orchil, generally described as orchella weed, were imported from the Canary Islands.

Orchil and litmus are known as indicator dyes, because they share the same behaviour, which is also characteristic of turnsole, turning blue in alkali and reverting to red in acid. Litmus is grouped with blue colour samples in de Mayerne's manuscript where its reversion to red is quite distinct (*Plate 8*). Colour-makers were familiar with the behaviour of those dyes long before it was described in detail by Boyle in the seventeenth century.[91] A traditional method for preparing a violet-blue from orchil is described by Peacham:[92]

> Korke or Orchall. Take fine Orchall and grinde it with vnslekt lime and vrine, it maketh a pure violet; by putting to more or lesse lime, you may make your violet light or deep as you will.

Litmus is the name used by Hilliard, who recommends that it should be used in the same way as indigo for shading other blues, and by Norgate, who suggests that it is unsuitable for use in limning. A defect of litmus blue was its tendency to revert to red, which is mentioned by Goeree:[93]

> Lakemus, because of its brown colour, serveth in nothing of its self alone, but only to some dark and rainie skies; and even in this it must be somwhat mingled with some other blew before, for otherwise it would turn red, and soon vanish . . . to hinder that, and to make a good blew out of the same, you must temper the Lakemus with good and clear Soap-boilers Lye; and if in standing it has lost its colour, by puting in it a little lime, you may make the same a great deal fairer than it was before.

Litmus is mentioned more frequently than turnsole in eighteenth-century books on water-colour painting, but, even so, it is doubtful if it was used extensively, because, according to *The Art of Drawing,* 1731, litmus was not supplied by colour shops and anyone who wanted to use it had to obtain it from a druggist or dry-salter and prepare it himself.

Logwood

The American redwood tree *Haemotoxylon campechianum* supplies a natural orange dye which may be turned violet or blue in the same way as the other indicator dyes mentioned above. Discovered in America in the sixteenth century, the wood was imported to Europe in the form of large blocks, hence the names logwood and blockwood. It was also called Campechy, Province or St Martin's wood. Logwood was used mainly in textile dyeing and its import to England was prohibited by Act of Parliament in 1581 on the grounds that it was a 'devilish drug' and a fugitive dye which injured cloth. The measure, which was intended to

Fig. 21. Logwood chips. (Reproduced by the kind permission of Messrs Winsor and Newton Limited.)

protect the native woad industry, affected painters very little, but a number of later writers on dyes and artists' colours have misinterpreted the statute and have suggested that painters were affected because indigo was prohibited.[94] Such was not the case, for indigo was imported and used in large quantities by the British in the early seventeenth century. The restriction did not altogether affect logwood either, because the early Stuarts granted letters patent for various people to import the wood and the statutes condemning it were repealed by 1662.[95]

Logwood is mentioned as a cheap and commonplace raw material in seventeenth-century works on water-colour painting. That is not to say that it was much used, for, although the liquid dye is suggested as a violet ink in B.M. MS. Sloane 3292, f. 4v, Norgate dismisses it as one of the colours fit only for washing prints. It is mentioned for that purpose by other writers of the same period, but none shows much enthusiasm, and a suggestion that it should be mixed with brasil red to make a transparent purple, which is included in *The Art of Drawing,* 1731, is one of the last references to logwood for painting purposes. There can be no doubt that its tendency to revert to red made it unsuitable for any kind of permanent work.

Blue Lake

This colour name appears in only two seventeenth-century English books on painting: in the English translation of Goeree's book on water-colour painting, and in B.M. MS. Harley 6376, where it is included obviously as an afterthought in Gyles' list of oil colours. The English painter might have adapted the name from Scheffer, a German author, who includes *Gummi Lacrae* in his list of blues, but it seems most likely that Gyles copied the name from Goeree. That writer describes the colour as being little different from indigo but more used in oil colour as it is less liable to fade; anyone wishing to use it in limning can proceed with it as if it were indigo.[96] Blue lake was certainly not a lake pigment in the modern sense, and it seems unlikely to have been a pigment which was prepared in the same way as indigo as Goeree suggests. Scheffer describes it as a gum lake which exudes from trees in the Indies.[97] His description suggests that it may have had some relationship with the *lacree,* used in dyeing and pharmacy, which is mentioned in early records of the British East India Company. Trees from which it might have been obtained are discussed in the printed edition of the Company's correspondence, but its identity is not definitely established.[98] There are no further references to blue lake as a pigment after the seventeenth century, although it is worth mentioning that Field has a sample of sap blue in his 'Practical Journal 1809', f. 318. He does not mention its origin but says that it is in hard lumps that can be used in water colour in the same way as gamboge or sap green, that is, the colour may simply be lifted by applying a wet brush. Apparently it was not permanent to light and it was unsuitable for use in oil. The fact that Field does not mention it in *Chromatography* suggests that he did not give it serious consideration as an artists' colour.

Cornflower Blue

Seventeenth-century writers were well aware of the fugitive nature of many of the organic colours which were available then, particularly of those made from flowers. It was not a painter but a scientist, Boyle, who recommended a blue dye prepared from cornflower petals, and, such was his reputation, that the colour was recommended for water-colour painting in some eighteenth-century books, sometimes under the name cornflower blue and sometimes as Boyle's blue or cyan blue. The colour was strongly criticised by Hoofnail, but mention of the colour in other books, largely those which derive a number of comments from *The Art of Drawing,* 1731, has left an unfortunate and probably erroneous impression that eighteenth-century painters in water colours were heedless of permanence. There is no evidence that cornflower blue was ever supplied commercially.

Woad

Woad, the pigment obtained from the leaves of *Isatis tinctoria,* has a long history of use and for centuries the plant was grown and the colour was prepared in England. It was comparable with indigo and was made in a somewhat similar way, but differentiation has always been made between woad and indigo.[99] Woad was virtually obsolete as an artists' colour by the beginning of the seventeenth century, whereas use of indigo increased.

A traditional method of obtaining the pigment for painting purposes was to collect the froth floating on top of a dyeing vat, preferably before the cloth to be dyed had been dipped in it. The same method could be used to obtain indigo, and in either case the colour was known as florey, that is, flower of woad or indigo. Instructions for mixing florey with a white base to make a lighter blue are included in a seventeenth-century manuscript amongst recipes transcribed from earlier sources:[100]

> To make blew Inde
> Take powder of Chalke and the flowre of a woade fat and faire Clene water and grind them on a stone and put it into a vessell and let it drie and then grinde it ofte in the same maner wise and let it dry iij times for the more it is done the better wilbe the Coullor and then make pellets therof and drye them in the *umbre* (shadowe). When thow wilt worke ther with grind it with Cole made of the Cuttings of lether and a littell water or ells temper it with oile and grind it with the same.

It is not always clear to the modern reader as to whether an early writer refers to indigo or woad, even though the writer himself doubtless knew the difference if only because woad was so much cheaper than indigo. In the passage quoted the colour is called Inde and there is no real indication as to whether the froth on the woad vat was made up of indigo, woad or a mixture of the two, although the context of the passage and its association with several other medieval recipes suggest that woad is a definite possibility. Peacham mentions *Indebaudias* and *English Indebaudias;* the latter was inferior, so presumably the English variety was woad. Writing in the late seventeenth century, John Smith says that some painters use blue balls instead of expensive indigo, thus implying that decorators might obtain woad balls as they were prepared at the woad-works. However, Smith does not mention woad by name. His is about the last comment which may be interpreted as a reference to woad as used in painting, but one must remember that Smith was writing of oil painting as decoration not as a fine art. No professional artist of the seventeenth century mentions woad, and it seems certain that indigo had then taken its place.

Indigo

The leaves of plants of the genus *Indigofera* contain a colourless glucoside which can be extracted and precipitated in the form of a dark blue pigment. Indian indigo was known from early times and it was imported to Europe by the overland route, hence the name *Indebaudias,* indigo from Baghdad, which appears in *Limming,* 1573, and a few later sources. Another name used occasionally in early seventeenth-century sources is *anil,* or the abbreviation *nil,* which is associated with the Arabic word for indigo; it was used not so much by painters but by merchants who traded in indigo.

Fig. 22. Indigo plant from Regnault, *La botanique*, 1774. (Reproduced by courtesy of the Royal Horticultural Society.)

As an article of commerce indigo played an important part in the early history of the East India Company. Spices from the East Indian islands were much more highly prized, but the Dutch dominated the spice islands and the British had to be content with second best in the form of the products of India. In the early seventeenth century trading stations were set up there, and the British factors traded without any thought of imperialism. At first they took no part in the production of indigo, although some reported on the procedure.[101]

> Indigo is made thus. In the prime June they sow it, which the rains bring up about the prime September; this they cut and it is called the Newty . . . a good sort. Next year it sprouts again in the prime August, which they cut and is the best Indigo called Jerry. Two months after it sprouts again, . . . thereof they make the worst sort; and afterwards they let it grow to seed and sow again. Being cut, they steep it 24 hours in a cistern of water; then they draw it into another cistern, where men beat it six hours forcibly with their hands till it become blue, mixing therewith a little oil; then having stood another day, they draw off the water and there resteth settled at the bottom pure indigo (which some to falsify mix with dirt and sand); which they dry by degrees, first in cloths til the water be sunk from it and it be curdled; afterwards they dry it in round gobbets.

Another account is included in *Pvrchas his Pilgrimes,* where the test for good indigo is explained. It consisted of placing lumps of indigo in water; good quality indigo floated, but pigment which was contaminated or of poor quality sank.[102] Sand and dirt sometimes contaminated indigo which was placed on the ground to dry, but it was often added deliberately. Indigo manufacture in the late seventeenth century is illustrated in *Figure 23*.

Adulteration of indigo was only one of the difficulties which the East India Company had to face. During the 1630s the Company had to meet increasing competition from quantities of good quality indigo imported from Guatemala by way of Spain, which was accompanied by a consequent drop in income derived from the Indian product. Reprimands of increasing severity were sent from London instructing the factors to stop adulteration with sand, to cease sending short weight and to pack the indigo in a more economical way. The factors had their own difficulties; they were not in charge of manufacture, competition from Asian merchants meant that they had to buy before the indigo had dried out completely (hence the apparent short weight), and timber for packing cases was practically non-existent so that they were forced to send indigo packed in bundles. An obvious solution was for the British to take charge of indigo manufacture themselves. The policy was adopted by 1649, and the British had taken a first step towards interference in Indian affairs.[103]

When the Company's indigo arrived at London it was garbled merely by sieving so that good lumps of rich indigo were separated from the dust, which was generally considered useless owing to contamination. It

Fig. 23. Indigo preparation in the French West Indies during the late seventeenth century: in the background, indigo plants are being cut; on the right, they are placed in the steeping vat; the beating apparatus is in operation; on the left, the pigment already collected in conical filter bags is carried to the drying shed (Pomet, *Histoire générale des drogues*, English translation, 1712).

does not seem to have occurred to anyone in the Company that pigment particles could have been separated from dirt by the simple measure of washing over, which was known to anyone who prepared artists' pigments. Such was the process probably employed by a London grocer who was granted letters patent in 1634 'to separate sand from flatt indigo and to make the fine pieces up into good indigo'. The East India Company's opposition to the enterprise is discussed in Appendix Three.

The best quality indigo was used by painters. Gyles refers to it correctly as rich indigo, but the term was corrupted in the eighteenth century and a number of writers refer to it as rock indigo or even stone blue. Painters in oils appear to have regarded indigo with mixed feelings. In the sixteenth-century manuscript, V. & A. MS. 86.EE.69, which contains information on oil painting, the colour is mentioned in the price list and elsewhere but not in the list of oil colours. De Mayerne provides conflicting information: that it is useless in oil as it fades, that it is no good unless varnish is used in conjunction with it, and that it may be given special treatment to prevent fading. Soaking in strong vinegar for two or three days is one suggestion, and soaking in urine in the sun for five or six days is another.[104] Marshall Smith suggests that if indigo is placed in a tightly stopped earthenware pot and boiled for four or five hours it will hold its colour in oil, or it may be buried for eight or ten years to achieve the same result. Indigo is seldom mentioned in the suggested colour mixtures which de Mayerne collected from various painters, but he mentions its use in providing a ground for glazing with ultramarine. Bate also refers to its use in under-painting, describing it as: 'A false blew. Blew of Inde is to make a false ground for a blew, and it must be ground with oil.' Referring to decorative painting, John Smith says that the oil colour is almost always reduced with white. Marshall Smith suggests that it should be ground in linseed oil, not nut oil as is recommended for other blues such as ultramarine and smalt. His is one of the last references to indigo in regular use in oil painting. Eighteenth-century writers exclude it completely, as Bardwell does, or they point out that it is used mainly in water-colour painting.

Limners were much more enthusiastic about indigo. It is mentioned by Hilliard and Peacham, and Norgate states that the colour 'ground finely workes sharpe and neate, it is of exceeding greate vse'. It can be mixed with bice and pink (greenish yellow) for sea green, and when mixed with white it is often used for flesh shadows in portrait minatures.[105] John Smith reiterates that the colour must be very finely ground and suggests that some alkali should be added at the same time to heighten its colour. Whereas indigo was replaced by Prussian blue in oil, it was retained by eighteenth-century and nineteenth-century landscape painters in water colours because it flowed very much better than Prussian blue. Varley points out that indigo is not really a bright enough colour for a sky on a clear day, but it is often used for the purpose because of its good working properties. He recommends mixing indigo with burnt sienna and gamboge to make a green for trees. Dagley also states that indigo is a good blue for landscapes and general use, and he describes it as 'tolerably permanent'. By the nineteenth century considerable care was devoted to its preparation, and Field cites intense blue as the colour name for indigo refined by solution and precipitation so that it is more permanent than the ordinary pigment. Consequently, indigo retained an important place as an artists'pigment, and, together with madder red, it may be counted as one of the organic colours with an exceptionally long history which extends from early times into the twentieth century.

Prussian Blue

Prussian blue, ferric ferrocyanide or similar compound, has been described as 'the first of the artificial pigments with a known history and an established date of preparation'.[106] However, the idea that a definite date may be associated with it may be a modern misconception. It is true that the circumstances of its accidental discovery were reported by Stahl some years after the event and that his account has been referred to many times since, but it is noticeable that throughout the period up to 1835 every writer who mentions any date in connection with the pigment states that it was discovered *circa* 1710, whereas the date, 1704, given by so many later writers was not mentioned until the late nineteenth century. Earlier writers based the date on the knowledge that the pigment had been advertised for sale in Berlin in 1710 and on the words of the German chemist, Stahl, whose account of the discovery was printed in 1731. He wrote of it as occurring by chance twenty years previously *'ante quatuor forte lustra'* (one *lustrum* being a period of five years). Unfortunately, nineteenth-century and twentieth-century writers who quote the date 1704 fail to state the source on which their information is based, and, in the absence of new evidence, any enquiry into the date of the discovery must begin with an examination of the remainder of Stahl's account: he mentions two people, Diesbach the colour-maker who made the discovery and Dippel the alchemist who supplied some of the raw materials, and he states that both were resident in Berlin. Dippel did not live there permanently, however; an eighteenth-century biography contains the information that just before 1704 he was in Giessen and Darmstadt, and, after some time, he went to Berlin. From there he went to Frankfurt-am-Main and then moved on to Holland at the end of 1707.[107] This evidence suggests that the discovery of Prussian blue could have been made at any time between 1704 and 1707. A modern authority states that Dippel is known to have been practising chemistry in Berlin in 1705.[108]

Stahl's account of Prussian blue was written because the fortuitous nature of its discovery appealed to him. The chance manufacture of the pigment resulted indirectly from Dippel's production of an animal oil which was distilled over some potash which was then treated as waste. Diesbach, who used to make Florence lake from cochineal, alum, English vitriol (ferrous sulphate) and a fixed alkali, ran short of alkali and asked Dippel for some of the potash which he saw had been thrown away. He was allowed to use it, and, after he had proceeded by his usual method, the lake appeared to be very pale. When he attempted to concentrate it, it turned purple and then deep blue. Diesbach returned to Dippel for an explanation and was told that the potash was tainted with animal matter. Stahl's account concludes with the comment that for some considerable time the pigment was made only in Berlin.[109]

Details of the manufacturing process were kept secret until 1724, when an account was sent from Germany to Woodward in England who allowed it to be published in *Philosophical Transactions.* The instructions were lengthy, but the method can be summarised as follows. To an alkali calcined with bullock's blood, dissolved and brought to boiling point, a solution of alum and ferrous sulphate was added while also boiling. During the effervescence which followed the mixture turned green, and, after it had been allowed to stand, it was strained. The residual greenish precipitate turned blue as soon as spirit of salt (hydrochloric acid) was poured on it. The pigment was then left to stand and was washed several times with pure water the next day, after which it was filtered and dried under gentle heat.[110] Woodward's communication was written in Latin and it doubtless became common knowledge quite quickly. By the 1730s manufacture of Prussian blue was widespread, as Shaw states in his *Chemical Lectures,* which contain manufacturing instructions in English: 'The Method of making this Prussian Blue in perfection, has been held and purchased as a very valuable Secret, both in England, Germany and elsewhere; but it is now got into several hands.'[111] No evidence has been found concerning the early manufacture of the pigment in England, nor has it been possible to verify statements concerning early manufacture which appear in nineteenth-century and twentieth-century works. For example, a number of modern writers give the impression that a colour-maker named Wilkinson was the first English manufacturer and that Wilkinson's blue became a synonym for Prussian blue. The information can be traced back to Hurst writing in the late nineteenth century, but he merely states that Wilkinson developed the pigment.[112] Literary sources of the eighteenth century contain no evidence that the name Wilkinson's blue was ever used by artists. Prussian blue is the name under which Berger was manufacturing and selling the pigment near London in 1766.[113] Colour names listed by Field in the nineteenth century are Berlin blue, Parisian blue and cyanide of iron. The origin of the first and last is obvious, and the second can be explained by the fact that a good quality blue was made in Paris. During the second half of the eighteenth century, considerable research was undertaken by French chemists in order to analyse the pigment and extend its use to the dyeing industry.[114]

Although Prussian blue was at one time used as a dye, it must be emphasised that it was originally advertised in 1710 as a pigment for artists' use. Following a summary of the limitations of other available blues, the notice in *Miscellanea Berolinensia* contained the announcement that the new blue which had been discovered a few years previously had been subjected to accurate tests. It was said to be absolutely durable in either oil or water colour and totally unaffected by nitric acid, fire or exposure to air. It could be ground to an impalpable powder and easily tempered with a knife, so it was suitable for miniature painters and oil painters alike; in addition, its softness meant that it would brush out well and mix easily with any other colour. Its versatility was such that, at full saturation, it was useful in painting shadows and, when thinned, it could be used as a lighter and brighter colour without any need for tinting with white. A great recommendation was its non-poisonous quality; it was said to be made from a kind of sugar so that it was edible, which meant that beginners could safely lick their brushes as they were liable to do. Finally, its price was attractive, being scarcely one-tenth that of ultramarine.[115]

Following the extravagant claims of the original manufacturer, one might expect the pigment to have been acclaimed immediately by artists, but such was not the case. Dossie states that anyone desiring permanent Prussian blue should prepare it himself instead of buying the pigment from a shop, because the commercial sort varied in strength and was unreliable. He further states that it can be used in all techniques except enamel, apparently overlooking fresco, a technique for which it is unsuitable. Quite possibly English painters were not particularly concerned with fresco, but Dayes mentions that Prussian blue is liable to be destroyed by alkali and that its colour is extracted by lime. Although no very early references to the uses of the oil colour in England have been found, it appears to have been well established by the middle of the eighteenth century, and painters in oils were well satisfied with the colour. Bardwell, in whose paintings the pigment has been identified, states that it 'is a very fine Blue and a kind working Colour', adding only

Fig. 24. A portrait by Thomas Bardwell in which Prussian blue has been used for the coat; it is of John Campbell, 2nd Duke of Argyll and Duke of Greenwich, dated 1740 (Reproduced by courtesy of the National Portrait Gallery).

the reservation that it should not be used alone in painting flesh. Nevertheless, he does not give it such high praise as ultramarine, and it was not until the second half of the eighteenth century that Prussian blue came to be the most important oil colour blue, as in Williams' *Mechanic of Oil Colours* where it is the only blue mentioned. The colour was reasonably priced, so amateurs used it as well as professionals, although not always without difficulty. It was probably no coincidence that a writer chose Prussian blue, a strongly staining colour, as an example in describing the difficulties to which self-taught amateurs were prone: '. . . a bladder of Prussian blue bursts over one's arm, and paints one's fingers and clothes.'[116]

In water-colour painting Prussian blue was held in distrust for a considerable time. It appears somewhat as an afterthought in Smith's *Method of Painting in Water Colours,* where ultramarine is recommended for the best painting, otherwise smalt 'or Prussian-Blue will do as well'. In *The Art of Drawing,* 1731, the colour is said to be difficult to use because of its oily quality. Water colours were kept in shells ready for use, and it appears that whenever a wet pencil was applied to Prussian blue in the shell the colour went yellow where the water ran round the edge, suggesting therefore that the pigment was poorly manufactured at that time. Naturally, the presence of yellow was unwelcome because it accentuated the tendency towards green which is a natural characteristic of Prussian blue. By the end of the eighteenth century its manufacture must have improved, because Payne states that it is a good colour for miniature painting and that no other blue can equal its strength and transparency. Even so, there were many complaints that it did not flow freely and some writers cast doubts on its permanence; in his fading tests Field found that the colour would fluctuate, fading on exposure and then regaining some of its colour after conclusion of the tests.

Antwerp Blue

The name for this colour, which is generally accepted as a pale variety of Prussian blue, suggests that the pigment was of foreign origin, but English sources provide no evidence on the subject, and, as there are numerous recipes for making the pigment, there is no reason to suppose that its manufacture was restricted to the Netherlands. An early reference to Antwerp blue is to be found in Payne's book on miniature painting, where it is described as a colour of recent introduction:[117]

> It certainly is one of the greatest deceptions in the world, being when dry a most beautiful bright Blue, but when wet and prepared a very dingy colour and totally unfit for the face of a miniature. It is, I imagine, a compound of Prussian Blue, Verditer, and some kind of White, as you will perceive on breaking a lump of it, white specks all through it. It may be used in blue draperies or back-grounds, but in nothing else.

Payne's supposition that Antwerp blue might contain a copper blue is confirmed by manufacturing instructions given by de Massoul, who suggests that either green copperas (ferrous sulphate) or vitriol of copper (copper sulphate) can be used. Field states merely that Antwerp blue is a weak variety of Prussian blue; obviously copper blue was omitted from the colour in the early nineteenth century.

In modern works Prussian blue and its weaker counterpart, Antwerp blue, are often grouped with inorganic pigments. However, when first introduced they were regarded as organic pigments owing to the fact that blood was one of the constituents used in the manufacture of Prussian blue. This led some eighteenth-century writers to suggest that the investigation of other organic substances might lead to the discovery of more artists' colours. Nevertheless, the idea was at that time unfounded, as it was the development of inorganic chemistry during the late eighteenth century which led to the introduction of new pigments.

6

Greens

Terre Verte: Fe, Mg, Al, K, hydrosilicate

The green minerals celadonite and glauconite, both generally known as terre verte, were used as a pigment in Europe during the seventeenth century, but English writers of the early seventeenth century make no mention of it, de Mayerne's reference in French to *bol vert* from Italy being an isolated example in the literary sources. It is mentioned later on, however, in the late seventeenth-century, anonymous MS. AL.41C at Sir John Soane's Museum, and by Marshall Smith, who states that a little drying oil should be used with it. The eighteenth-century customs records support the view that the pigment was not extensively used in England, but it obviously grew in popularity at that time and it is the only green listed by Bardwell. Dossie describes it as a blue-green pigment which is not very bright, is semi-transparent in oil, but has strong body in water colour. He suggests that it owes its green colour to copper, and a

Fig. 25. Malachite, *left*, and lumps of green earth, *right*. Early writers were usually careful to distinguish between earth colours and pigments obtained from semi-precious stones, a much harder raw material. (Reproduced by the kind permission of Messrs Winsor and Newton Limited.)

number of writers of the late eighteenth century also make the same suggestion, but Field rightly points out that green minerals containing copper, which were sometimes called terre verte, are not the same as true terre verte, which contains iron. Some sources of the pigment which he lists are Italy (Verona), France and Cyprus.

Chrysocolla: $CuSiO_3 \cdot nH_2O$

The colour name ceder green, was used by sixteenth-century and seventeenth-century painters in water colours to refer to an inorganic green pigment which was almost certainly chrysocolla (copper silicate). It must be admitted that no evidence for use of chrysocolla as an artists' colour has been found by those engaged in the analysis of pigments in paintings and that some are doubtful if it was ever so used. However, documentary evidence from the late sixteenth and early seventeenth centuries indicates that a natural inorganic green other than malachite was favoured as a water colour, and its restriction to water-colour painting would therefore make identification unlikely as analysis of pigments has to date largely been confined to easel paintings. Since documentary evidence in several languages point to chrysocolla as the most likely identification of this water-colour pigment, it is here described under that name.

The pigment was once used as an adhesive for gold; that is the literal meaning of *chrysocolla*, which is derived from Greek χϱυσος (gold) and κoλλα (glue). The English name *gold solder* was used for the same substance during the medieval period, and the name *ceder* is probably a corruption of solder.[118] Ceder is the spelling used by Hilliard and Norgate and in V. & A. MS. 86.EE.69. Other slightly later writers, including Bate, Sanderson, Goeree and Gyles, adopt variations in spelling, and it is doubtful if any of them were familiar with the pigment, for ceder green became obsolete during the early seventeenth century.

None of the English sources supplies a detailed description of ceder green, although they contain an implication that the pigment was inorganic. There might be some doubt that it was malachite—basic copper carbonate—and not chrysocolla, for Hilliard describes ceder green as the best green for limning and does not mention any of the known names for malachite. Nevertheless, green bice, the seventeenth-century name for malachite, is listed in addition to ceder green by Norgate and in B.M. MS. Sloane 6284, so it is clear that ceder green and green bice were not synonymous. Amongst the water-colour samples in B.M. MS. Sloane 2052, de Mayerne gives *Crisocolla* and *Berggrün* for one sample and includes *Lasürgrün* as a separate sample (*Plate* 7). In Goeree's book, *bergh green* (that is, mountain green), probably malachite, is mentioned as a distinct colour from *sedergreen*. Inclusion of the colour in Goeree's work enables the name to be traced in other languages, for it appears as *sever-groen* in the Dutch and as *Seifer-grün* in the German translation. The pigment can be traced in German sources of the sixteenth century, one of which is the *Illuminirbuch* by Boltz, where a description of *Schifergrün* follows that of *Berggrün.* Both were blue-green mineral pigments, but the author considered *Schifergrün* the better of the two:[119]

> Schifergrün wirdt auss den Ertzsteinen gemacht, eines schöner denn das ander, ist ein schwere Materei wie Lasur .... Das Schifergrün so es schön ist, so muss mans nicht fast reiben, es verleuret sonst sein schöne grünfarb von dem hart reiben.

The author asserted that too much grinding would spoil its fine colour. That point, which is applicable to several mineral pigments, was implied by Norgate when he listed ceder green amongst the pigments which should be washed but not ground.

Another sixteenth-century German source of information is Agricola who in a Latin–German glossary included *chrysocolla: Berggrün und Schifergrün.* Elsewhere the same writer explained how water was able to wash away deposits of chrysocolla in the Carpathians and the Harz mountains, and he both illustrated (*Figure 26*) and described the series of tanks on different levels in which the pigment settled while the water

Fig. 26. Chrysocolla collection from a mountain stream by the method used in the Carpathians and Harz mountains and described by Agricola during the sixteenth century: A is the mouth of the tunnel through which the stream bearing copper deposits emerged; B is the channel which directed the stream down an incline so that the green pigment could settle in the tanks C (Agricola, *De re metallica*, 1556).

flowed down an incline from one tank to another. The green pigment deposited in that way was then collected and sold.[120]

It is somewhat surprising that, as water-colour painters held chrysocolla in such high esteem during the late sixteenth century, it was mentioned so seldom in the seventeenth century and was virtually unknown subsequently. It is possible that there was some disruption of trade and that chrysocolla became difficult to obtain in the same way that azurite, and to some extent malachite, became scarce. At a later date, efforts to obtain azurite met with success, but, by that time, chrysocolla had been forgotten, as another traditional pigment, terre verte, had become well established and a variety of manufactured copper pigments had been introduced.

Malachite: $CuCo_3 \cdot Cu(OH)_2$

Another naturally occurring copper green, malachite, is to be found in association with both chrysocolla and azurite.[121] Malachite and azurite are basic copper carbonates which were called green and blue bice in England during the seventeenth century. However, the green pigment was recommended less frequently than its blue counterpart in written sources. Although it is described in B.M. MS. Sloane 3292, f. 3v, as 'the pewrest greene for Lymming', it is not mentioned by Hilliard, and Norgate states that it is too coarse for many purposes—yet another reference to the fact that, if finely ground, some pigments become too pale to be of any real use. Its coarseness doubtless accounts for its total absence from lists of colours recommended for water-colour painting in the second half of the seventeenth century. During the following century the name green bice was sometimes interpreted as meaning a mixture of a yellow (orpiment) and of blue bice (which was by that time a name for smalt), a statement which appears in *Dictionarium Polygraphicum* and several books of later date; however, the genuine pigment was still available at the end of the eighteenth century, as it is listed as malachite or mountain green by de Massoul. The same author also lists azure green, which may be the same pigment because the name is associated with azurite and duplication of names is a feature of that colourman's list. Although Field mentions mountain green, citing malachite and green bice as subsidiary names, the pigment was displaced during the nineteenth century by manufactured greens of superior brilliance.

Green Verditer: $CuCO_3 \cdot Cu(OH)_2$

Literature concerning manufactured green basic copper carbonate is not very extensive although it has been identified in easel paintings of the fifteenth century and one may assume, therefore, that it was used more than has previously been supposed.[122] The pigment could be made by the same process as blue verditer: as by-products of silver refining, verditers were made by pouring copper nitrate on whiting, after which the pigment was drained, washed and dried. In the seventeenth century manufacturers found it much easier to make green verditer than blue, so the green pigment was relatively cheap.

Little can be said of verditer as an artists' colour, because, according to English documentary sources, it was seldom used for the finest work. It is true that the colour is included in the price list in the late sixteenth-century V. & A. MS. 86.EE.69 and that Hilliard's treatise includes the remark that verditer may be used if ceder green is unobtainable, but apart from several seventeenth-century references to Hilliard's use of the colour, all bearing a clear implication that if such an eminent limner used verditer the colour could not be wholly bad, there are hardly any references to its use in fine painting. Green verditer is likely to be the colour referred to by the name *verdet* in B.M. MS. Sloane 3292, f. 3v, where it is described as 'the roughest colure that belongeth to Lymming'. The application of the name *verdet* to verditer is unusual, because it was commonly used as an alternative name for verdigris, another green pigment. Nevertheless, it is not applied to that colour in this context, for verdigris is mentioned by name and praised on the same page. The reference to the roughness of verdet suggests that verditer, which was somewhat coarse and gritty, is intended.

By the eighteenth century, green verditer had disappeared from the artists' palette and it is not even mentioned by Dossie. In the nineteenth century, it was not recommended for fine painting, but the pigment was available, as, according to Field, it was sometimes sold under the name green bice.

Verdigris: $Cu(CH_3{\cdot}COO)_2{\cdot}2Cu(OH)_2$

Basic copper acetate, the oldest of manufactured copper greens, has been known since ancient times, hence the common name *verdigris,* which means green of Greece.[123] Other names were sometimes used in the past; Spanish green appears in *Limming,* 1573, and the Latin name *viride aeris,* literally copper green, appears in several English sources. As already mentioned in connection with verditer, verdet, which is a diminutive of the French word for green, was usually associated with verdigris, but, although it appears frequently in de Mayerne's notes, which were written in French, and sometimes in English books written under French influence, the name was never naturalised.

Instructions for making verdigris exist in manuscripts dating from the sixteenth century and earlier, but there are few instructions in seventeenth-century works because the pigment was then produced commercially. Montpellier in France was a noted centre for the manufacture of verdigris, and two detailed accounts of the procedure followed there were written in the seventeenth century, one in French by de Mayerne and another in English by Ray, a Fellow of the Royal Society who visited Montpellier during the course of an extensive foreign tour.[124]

According to Ray, a quantity of sour red wine was poured into large earthenware pots to a depth of two to three inches. Then grape stalks, previously soaked in wine, and small copper plates were placed in alternate layers on a grille above the level of the wine in each of the pots. Sufficient space was left between the plates and in the centre of the structure to allow acid fumes to spread through the contents. The pots were covered and left in a cellar for five or six days, after which they were opened, the copper plates were turned and replaced, and the pots left for a further three or four days. Next, the plates were removed, piled in small groups, sprinkled with the same vinegar-like wine and left for a similar time, after which they were pressed under weights for four to five days, and, finally, the green matter was scraped off the plates, moulded into lumps with the help of a little wine and then dried ready for sale.

Remarks of the few painters who recommend verdigris suggest that a certain amount of dirt had to be removed from the ordinary pigment before it could be prepared for use. A much better pigment was the refined variety, neutral copper acetate, known as distilled verdigris, which was prepared by dissolving crude verdigris in distilled vinegar, then filtering and evaporating the liquid to recrystallise the pure product. Neutral verdigris is less prone to colour change from blue-green to green than the basic variety, and during the seventeenth and eighteenth centuries distilled verdigris was for sale in the form of crystals clustered around small sticks. Nevertheless, it was comparatively expensive and most painters considered it too costly to be worth using even though it was superior to the crude variety.

Verdigris is recommended mainly by early writers, such as the authors of MS. Sloane 3292 and *Limming,* 1573, whereas most seventeenth-century writers offer numerous warnings concerning its lack of durability and its incompatibility with a number of other pigments. It was a quick drier in oils, and one of its main functions in oil painting was to act as a drier when mixed with black. Otherwise it was mixed with resin and used for glazing, a technique which is known to have been used by Flemish painters at an earlier period. De Mayerne noted a method for making a very transparent glazing colour (usually referred to today as green copper resinate), suggesting that verdigris and Venice turpentine should be mixed together and heated.[125] A recipe in *Academia Italica* suggests that well-ground verdigris and varnish should be mixed cold and thinned with turpentine if necessary. A tradition that a bright green could be obtained in oil painting by laying a glaze of copper resinate over a white ground appears to have continued from the seventeenth century until the end of the eighteenth century, when it was recommended in the *Practical Treatise*, 1795.

Neither Hilliard nor Norgate was in favour of the use of verdigris in miniature painting. It was more commonly used for washing prints during the second half of the seventeenth century, but it was then generally dissolved in acid and used as a clear, transparent liquid which was not known as verdigris but simply as copper green. *Academia Italica* contains a recipe for its preparation from one pound of copper dust, one

320. Specimens. &c.

Crystalized or Distilled Verdigris. of Sills &Co
1/4 lb. Oz.. From 40 to 50£ pr. Cwt. In crystals.
externally powdery. acetic smell. Chemically termed
Acetite of Copper. grinds frothy in gum water & works
so from the pencil. Whitened by a few days exposure
to the damp & mephites of a Jakes. & the same by
a few days's exposure to the Sun. & subsequently
blackened by further exposure to the mephites of the
Jakes. the same after 14 days. In Oil. Pasty.—
cerulean green.—heavy transparency.—teints pure &
Turquoise toned.—dried firmly some hours before the Oil.

Mineral Green. of Whitaker Bristol. pr.
Mr Middleton has it at 2/10
lb. 5/. In lumps of close texture. very absorbent &
upon wetting turn pale. smell earthy. rather
pasty. darkened when exposed a week to the mephites of a Jakes. rather bleached by 14 days sun in
Aug. Blackened by longer exposure to the Jakes.
In Oil. absorbent.—pasty.—pure.—yellow toned.
heavily transparent. not powerful. dried firmly before the oil, and its teints.

Deep Mineral Green. Of Hare Bristol
lb. 4/3. In lumps of close texture. very absorbent
turns pale on wetting. pasty. darkened when exposed a week to the mephites of a Jakes. rather
weakened by 14 days Sun in Aug. Blackened
by longer exposure to the Jakes. In Oil. as of
above, rather superior in depth & transparency
& inferior in colour. dried firmly before the oil
and before its teints. Very slightly affected by 5 weeks exposure
to the Sun & a Jakes.

Fig. 27. Field's comments on distilled verdigris, and two other copper green pigments from his 'Practical Journal 1809' (Courtauld Institute of Art, MS. Field/6, f.320). (Reproduced by courtesy of the Courtauld Institute of Art.)

pound of tartar and two quarts of water, with instructions that they should be boiled until half has evaporated; after the remainder has been allowed to stand, the clear green part should be poured off and used as a wash. The third edition of John Smith's book contains a recipe in which the ingredients recommended are one pound of verdigris and three ounces of cream of tartar. Copper green was reputed to be a good grass colour, whereas verdigris often required the addition of yellow in order to correct a tendency towards blue.

Until the end of the eighteenth century Montpellier was the source usually mentioned in connection with verdigris, and most authorities suggest that the best came from there. However, much was probably used which was not French at all, for the customs records of the eighteenth century show that quantities imported from the Netherlands and Italy usually exceeded that imported from France. At the same time it was being made in England, as a patent for its manufacture was granted in 1691 and the premium offered in the eighteenth century by the Society of Arts to manufacturers who could succeed in making the pigment equal in quality to the French, was awarded several times.[126] Throughout the whole of the period under discussion, crude verdigris was made on a large scale, because it was used in decorating and in the dyeing industry. Consequently, it was both cheap and easy to obtain, and, on that account, it was probably used to some extent in artistic painting throughout the seventeenth and eighteenth centuries despite the fact that it was seldom recommended by professional painters.

Scheele's Green: $CuHAsO_3$

In 1775 another manufactured copper green was discovered by the Swedish chemist, Scheele, as a result of his investigation into the nature of arsenic. The pigment, copper arsenite, is mentioned in an account of his experiments.[127] The paper does not contain detailed instructions for making the pigment, but its composition was made known because, as Scheele explained in 1777 in a letter to another scientist, he felt that potential users should be warned of its highly poisonous nature, and, in addition, he wished to prevent anyone else claiming credit for the discovery, an injustice which Scheele had suffered previously.[128] It was not until the following year that, at the request of the Stockholm Academy of Sciences, Scheele published detailed instructions for making the pigment. A number of authorities give 1778 as the date of discovery, but in his paper Scheele states quite definitely that the colour had proved itself over the past three years. To make the green, some potash and pulverised 'white arsenic' (that is arsenious oxide) were dissolved in water and heated, and the alkaline solution was added, a little at a time because of effervescence, to a warm solution of copper sulphate. The mixture was allowed to stand so that the green precipitate could settle, the liquid was poured off, and the pigment was then washed and dried under gentle heat.[129]

Fig. 28. Carl Wilhelm Scheele (1742–1786), from a posthumous portrait by Falander. (Reproduced by courtesy of the Ann Ronan Science Library.)

The new copper green was probably well known within a short space of time, for there was considerable exchange of scientific ideas by the end of the eighteenth century. It is mentioned, with a brief outline of how to make it, in the *Practical Treatise,* 1795, and, in a general discussion concerning artists' colours, Thenard pronounced that the only quality it lacked was a greater degree of intensity.[130] In 1812 a process for manufacturing it and methods for varying the colour were patented in England, (printed series number 3594). There must be some doubt about the extent to which Scheele's green was used in artistic painting, because it is not mentioned in any other English literary source of the period with the exception of Field's *Chromatography,* in which it is described as a light, warm, opaque green which, with Schweinfurt green, is superior in durability to other copper greens. Field includes a sample of Scheele's green that was obtained in Paris, in his 'Practical Journal 1809', f. 372, and in *Chromatography* contributes the additional information that an olive-coloured pigment can be obtained by burning either verdigris or Scheele's green.

Emerald Green: $Cu(CH_3{\cdot}COO)_2{\cdot}3Cu(AsO_2)_2$

Schweinfurt green, or emerald green as it is generally known in England, was introduced as a result of efforts to improve copper arsenite, but its development has received much less attention than that of Scheele's green, largely owing to lack of documentation.[131] By tradition, the pigment was first produced commercially by Russ and Sattler in Schweinfurt in 1814. An interval of eight years followed before a method of making copper aceto-arsenite was published; then in 1822 Braconnot and Liebig separately published papers on the subject. A summary of their methods with details of quantities is supplied by Mérimée. By Liebig's process, verdigris was dissolved in vinegar while warm and an aqueous solution of white arsenic was added to it so that a dirty green precipitate was formed. In order to obtain the correct colour, a fresh quantity of vinegar was added to dissolve the precipitate, and, when the solution was boiled, a new, bright green precipitate was formed. It was then separated from the liquid, washed and dried in the usual way.[132]

The improved pigment was more durable and had a more intense colour than other copper greens, 'this tribe of colours' as Field calls them, but its use was limited. The same writer states that it is a colour seldom encountered in nature although it is suitable for imitating the colour of gems. He also points out that it has the same tendency to blacken as all other copper greens. A colour sample of emerald green is included in Fielding's *Index of Mixed Tints.*

Brunswick Green

In the *Practical Treatise,* 1795, Brunswick green is described as 'a very valuable and newly discovered colour' prepared at Brunswick by brothers named Gravenhorst. The author was not familiar with details of its composition, although he knew that it was a copper green and he was able to say that a similar colour was sold by Brandram and Company of London. According to Chaptal, the pigment was made by covering copper filings with a solution of ammonium chloride and leaving them in a closed container. After the green precipitate was washed and dried, it was used in oil painting and for printing. This description agrees with the generally accepted views that the original Brunswick green was a basic copper chloride.[133] However, it seems that a mixed green was sometimes sold under the name Brunswick green in the nineteenth century, as Field differentiates between a colour commonly called Brunswick green, which he says is a mixture of chrome yellow and Prussian blue, and the genuine pigment, which he describes as a copper colour.

Cobalt Green: $CoO \cdot nZnO$

In 1780 Rinman, a Swedish chemist, published details of a new green pigment in the journal of the Stockholm Academy of Sciences.[134] Church states that Rinman made the compound of oxides of zinc and cobalt by 'precipitating with an alkaline carbonate a mixture of the nitrates of cobalt and of zinc, and then strongly heating (after washing) the precipitate formed'.[135] Most authorities suggest that cobalt green, otherwise known as zinc or Rinman's green, was not commercially available until the middle of the nineteenth century, when zinc oxide was manufactured on a large scale. Such a view may need modification, because there is evidence, which is discussed in connection with zinc white, pointing to the fact that zinc oxide was available in England and France at the end of the eighteenth century. It is true that cobalt green is not mentioned by writers of the late eighteenth and early nineteenth centuries, and, when it is mentioned at a slightly later period, Field states that the pigment known as cobalt green is often a mixture of cobalt blue and chrome yellow. On the other hand, Field gives such a detailed description of true cobalt green—a pure but not very powerful green which is a good drier and is durable in water and oil—that one is led to suppose that he actually made it, although there is no evidence of it in his 'Practical Journal 1809', a notebook he used until the early 1820s.

Oxide of Chromium: Cr_2O_3

The discovery of chromium in 1797 and the isolation of the element in the following year are discussed in connection with chrome yellow, because they resulted from research concerning the nature of a mineral which was occasionally used as an orange-yellow pigment. Vauquelin's paper of 1809 includes an account of his method of preparing the oxide from potassium chromate, and it also contains the information that the pigment was already being used in the Sèvres porcelain factory and in another at Limoges.[136] Oxide of chromium is an extremely reliable pigment which, but for its opacity and lack of brilliance, might have been adopted in painting as early as in the ceramic industry. Its dull, opaque quality doubtless delayed its introduction as an artists' colour, but the pigment did not remain unrecognised for as long as some writers suggest. Field states that, although the colour commonly known as chrome green is a mixture of chrome yellow and blue, there is a true chrome green which is very durable and can be made in various degrees of transparency or opacity so that it is suitable for painting in oils or water colours. The proof that he actually made the pigment is contained in the first edition of *Chromatography,* as he states that the sample of green included in his Plate I, figure 3 is oxide of chromium. It is fairly transparent, and, whereas it has neither brilliance nor the characteristic granulation of modern transparent oxide of chromium, it does not have the dull, dense appearance typical of the opaque variety. Another example of his own manufacture is included in his 'Practical Journal 1809', f. 358, in an entry dating from after 1815.

It has sometimes been supposed that neither opaque nor transparent oxide of chromium was supplied to painters until the second half of the nineteenth century, although it is now clear that painters who were supplied by Field could have obtained the pigment, and Pannetier, a French colour-maker, developed a transparent variety, hydrated oxide of chromium ($Cr_2O_3 \cdot 2H_2O$), now generally known as viridian, during the first half of the nineteenth century.[137] In the 1840s Winsor & Newton listed green oxide of chromium, which was presumably the anhydrous, opaque variety. Transparent oxide of chromium is included as well amongst oil colours in a catalogue dated 1849.

Pansy Green and Lily Green

Turning to organic colours, one encounters relatively few organic greens, far fewer than in the range of blue pigments or dyes. A green pigment was occasionally prepared from flower petals which were bruised and soaked in water to extract a dye which was then precipitated on alum. Pansy green is mentioned by Hilliard, and a recipe for making pansy green from the petals of violets appears in B.M. MS. Additional 23080, f. 26. *Flowre de luce* green is listed in MS. Sloane 6284 and the English name, lily green, is mentioned in *The Art of Painting in Miniature,* 1730. A suggestion that the petals of irises or lilies can be used for making sap green is contained in *The Art of Drawing,* 1731. Both lily and pansy green may be regarded as belonging to an earlier period, as writers of the seventeenth century mention them less frequently than sap green made from buckthorn berries.

Sap Green

Greens made from a variety of organic materials were sometimes called sap green, but in the seventeenth century the name usually indicated the colour made from ripe buckthorn berries, a variety of *Rhamnus*. In Gerard's herbal the particular variety is described as *Rhamnus solutivus,* the unripe berries of which are used to make yellow and the ripe berries, gathered in the autumn, are used to make green. There is no mention of adjusting the mordant to alter the colour from yellow to green; seventeenth-century practice was to prepare the pigment by steeping the berries, boiling them and adding alum. The recipe which follows comes from an early seventeenth-century manuscript:[138]

> To make Sapgrene. Take and gather a berry which is called wyrethorne at St. James Hyde, it groweth lyke sloes with leaves lyke a slotree. (Glovers doe dye skinnes therewith, and taffetaes are dyed therewith.) ffirst you must bruze them in cleane water for swellinge but with a lytle so as the water cover the berryes tis sufficient. Then let them boyle but one wallom or two.* Then scomme them and put into an earthen pott, then cast thereuppon a good quantitye of Allomepowder stirringe it well together, and keep it verie close. And yf it be drye when you would worke therewith laye a pece of yt in gomwater. And that is to dyaper, deepe or damaske vppon grene byce, verditur or verdigrece after the fflemish sorte.

The writer implies that, if air is excluded from the colour, it remains soft for some time, but he omits to mention the fact that it was often stored in bladders. Such containers gave rise to the French name *vert de vessie* and to the English *bladder green,* which is to be found occasionally in eighteenth-century English books translated or adapted from French works. The early seventeenth-century writer quoted above recommends the addition of gum water to sap green, but the use of gum was exceptional. Gyles states that the best sap green is as soft as tar and that it can be used immediately with the addition of water. However, it had generally hardened by the time a painter came to use it, and most books contain instructions for it to be soaked in water or vinegar to prepare it for use.

Sap green is frequently mentioned by painters in water colours, but, even though it was the only yellow-green they had ready for use without mixing, it was not a popular colour in the seventeenth century. Norgate states that it is so thin and transparent that it is unserviceable for most uses. In de Mayerne's manuscript the sap green colour samples have gone brown (*Plate* 7). Browne points out that the colour is omitted from the colour list in the second edition of his book on the grounds that it shines and fades. Criticism of the colour is also contained in *A Book of Drawing,* 1666:[139]

> Sap green is a dark dirty green and never used but to shadow other greens in the darkest places, or else to lay upon some dark ground, behind a picture, which requires to be coloured with a dark green, but

***Wallom* is the equivalent of *wallop* — a series of noisy bubbling motions made by water, or, one such bubbling motion used as a vague measure, in cooking, of the time anything is allowed to boil. (O.E.D.)

> you may make a shift well enough without this greene, for Indico and Yellow-berries make just such another colour.

As with a number of other extremely transparent colours, sap green gained in popularity during the eighteenth century. In *Bowles's Art of Painting in Water-Colours* it is described as the 'best green for Water-Colours our Age affords', and Payne recommends it as a useful colour in mixtures—adding, surprisingly, that the colour does not change. In the early nineteenth century it was once again condemned as an unsatisfactory colour, as Field states that it is ineligible in water colour and absolutely useless in oil.

Prussian Green

True Prussian green was made by the same process as Prussian blue, with the omission of the spirit of salt (hydrochloric acid) which turned the substance from green to blue. According to Dossie, the green pigment made in that way had gained some popularity in the early part of the eighteenth century but by the middle of the century it had fallen into disuse. A similar colour could be obtained quite easily by mixing Prussian blue and pink (yellow pigment), and it seems likely that the Prussian green sold at one shilling per ounce at the end of the eighteenth century was a mixture. Both de Massoul and Field mention it as a mixed green, so genuine Prussian green must be regarded as an extremely insignificant artists' colour.

7

Inorganic Yellows

Ochres: mainly $Fe_2O_3 \cdot H_2O$ or Fe_2O_3

The iron oxides called ochres, which vary from a dull yellow to red and brown, are outstanding for their long history of use as pigments; one shade or another occurs in surface deposits in many parts of the world, with the result that they have always been relatively cheap and easy to obtain. The long history of the pigment may account for the survival in sixteenth-century and seventeenth-century English sources of some obscure colour names, the origin of which is unknown. One such name is *oker de luke* or *luce* which appears in *Limming,* 1573, and is mentioned by Peacham. Apart from the information that it is a yellow ochre suitable for use in painting hair, there are no details about it, and Peacham, who often explains the derivation of colour names, offers no comment. Another name of obscure origin is *oker de rouse,* which occurs more frequently than *oker de luke* in seventeenth-century sources. It was likewise used as a hair colour, and Hilliard suggests that it should be used as a shading yellow if there is nothing better available, thus indicating that it was a dark ochre tending to red or brown. English oker de rouse may be compared with the French *ocre de ru* (otherwise *rut* or *ruë*), the meaning of which is equally obscure. In France in the seventeenth century it was suggested that ochre of that name came from Ruë, which was though to be an English place name.[140] The French name remained in current use longer than the English name; in the nineteenth century Mérimée suggested that *ru* was an old form of *ruisseau* and that ochre of that name once indicated ochre from a brook-side deposit.[141]

The place of origin was sometimes incorporated in the name of an ochre; *spruce oker* may be taken as an example, spruce being an old form of Prussia or Prussian. English sources indicate that it was a dull yellow. In B.M. MS. Stowe 680 it is listed as a faint yellow, and it is described as yellow in *Academia Italica* and other books printed during the second half of the seventeenth century. John Smith points out that it is darker than English yellow ochre.[142] A few English writers of the seventeenth century speak of English ochre, one of the first to do so being Norgate, who was particularly ready to praise a native product. English ochre was synonymous with yellow ochre, and John Smith remarks that most comes from Shotover Hills near Oxford, indicated by the name Oxford ochre, which was introduced in the nineteenth century. That deposit was worked until the mid-twentieth century. Gyles states that Bury ochre, which is better than the normal yellow, comes from 'St Edmunds Bury' in Suffolk. The occurrence of ochre in England is not restricted to Oxford, but Gyles was most likely mistaken about Bury St Edmunds, for the ochre he had used was probably Berry ochre from Vierzon in France, which was well known in Roman times and was mined until the nineteenth century.[143]

Italy is another source of natural ochre, but it is mentioned infrequently by English writers of the seventeenth century. Apart from de Mayerne's reference to it, it is mentioned only by Alexander Browne and Marshall Smith, both of whom include Roman ochre with the names of other ochres. It is not mentioned extensively during the eighteenth century, although it is described by Payne as a reddish yellow with good body and a warm tone. The mention of a tendency towards red agrees with de Mayerne's description of Italian ochre as a brown red, but nineteenth-century writers omit reference to a red tendency. Varley mentions that it is less bright than yellow ochre but is used for the same purpose, and Field places it between Oxford yellow ochre and dark varieties such as spruce ochre and *ochre de ru*. The particularly transparent variety of ochre which is found near Sienna and is known by that name is mentioned in English sources from the mid-eighteenth century onwards. Initially it was described as *Terra di Siena,* unburnt or burnt; not until the nineteenth century did the anglicised and abbreviated forms raw sienna and burnt sienna become common. The scarcity of references to Italian ochres in English documents suggests that they were little used in England during the seventeenth and early eighteenth centuries, a view which is supported by evidence of the customs records. In the ledgers examined for the first half of the eighteenth century, the only entry for ochre from Italy is that for one and a quarter barrels in 1730. During the same period greater quantities were imported from Holland, an average annual total being thirty barrels. Examination of ledgers for the second half of the eighteenth century shows a steady increase in the quantity of ochre imported from Holland, rising from 186 barrels in 1760 to 324 in 1780, but there is no evidence of corresponding increase in the import of ochre from Italy which references in literary sources would indicate. Painters' references to Italian ochre and sienna increased after the 1750s, thus indicating that British painters became much more familiar with the Italian product after artists such as Ramsay and Reynolds had visited Italy. The customs records confirm the fact that sienna was used on a considerable scale during the nineteenth century, as the import of several tons is recorded for 1820.

Literary sources show that native ochres were generally regarded as useful and reliable pigments. In regard to the preparation of yellow ochre, de Mayerne warns that it should be ground only in oil and not in water, which retards its drying time and causes it to lose colour. Norgate too recommends that it should be ground and not washed. All the writers on oil painting who distinguish between essential colours, and those which are merely useful, place yellow ochre in the first category, and Bardwell refers to light ochre as a 'friendly mixing colour, and of great use in the Flesh'. On the other hand, painters in water colours tended to differ in opinion concerning yellow ochre. Whereas Hilliard suggests that it should be used if there is no better yellow available, Norgate states that it works well when finely ground and that it is useful for a number of purposes such as hair, drapery and buildings. Its lack of transparency led some eighteenth-century writers to omit yellow ochre from the selection of colours suitable for washing prints, and, for the same reason, Varley suggests that it should not be used in foliage, although its use for buildings is acceptable. One would expect the transparency of raw sienna to be popular with painters in water colours, but Payne states that the colour is greasy and inferior to gallstone. Nevertheless, an increasing number of painters used the pigment, and, according to the evidence of Henderson, it was useful as a shading colour in nineteenth-century flower paintings. Yellow ochre does not appear to have been much liable to adulteration, although Dossie warns that Dutch pink is sometimes mixed with it and that such adulteration is apparent if the pigment is heated. In addition, Williams claims that it is difficult to obtain the best genuine ochre. In view of both remarks, it seems likely that during the eighteenth century some unscrupulous tradesmen may have attempted to brighten the pigment, which is a naturally dull colour, but complaints concerning the practice are rare and it seems that most painters were satisfied with the colour they purchased, which was, after all, so cheap that it was hardly worth falsification.

Manufactured Iron Oxide—Mars Yellow: $Fe_2O_3.H_2O$

The name Mars yellow is a late eighteenth-century, and therefore modern, variation of the Latin name *crocus martis,* which was used from a relatively early period by alchemists to indicate yellow iron oxide. Crocus was associated with the yellow pigment, saffron, and Mars was an internationally recognised name for iron; hence the name literally meant yellow of iron. The close association of ideas between crocus and saffron is demonstrated by de Mayerne and Atkinson, an eighteenth-century patentee, both of whom used the words interchangeably; in each case *crocus martis* appears as the principal name with saffron as a secondary title.

In view of the fact that alchemists made iron oxide, it is strange to find that it is scarcely mentioned in seventeenth-century sources. De Mayerne refers to it, but the main object of his discussion is to point out methods of converting the yellow substance to red for use in enamelling.[144] It is possible that artists applied the name *ochre* to all iron oxides, natural or manufactured, but this is improbable because, as many documents include discussion on the sources and qualities of various native earths, some mention of an artificial ochre would have been likely had it been used. The conclusion must be that a plentiful supply of natural iron oxides existed, so making the commercial production of a manufactured equivalent unnecessary. Documentary sources suggest that Mars yellow did not come into general use in England until the late eighteenth century, when it probably became available as a result of the expansion of the chemical industry, not because of a demand on the part of painters.

Eighteenth-century sources do not contain a really satisfactory explanation of how the pigment was made, but the method was probably little different from that used in the nineteenth century. Mars yellow was made from ferrous sulphate, which was known in the past by other names such as green vitriol or martial vitriol. The solution was mixed with alum and precipitated by means of an alkali, such as caustic soda, potash or lime, and upon exposure to air the precipitate turned yellow.[145] An early reference to it, although as *oker* not as Mars yellow, appears in an extremely complicated and obscure patent specification dated 1780, and a further reference, this time to '*crocus martis* or saffron of Mars', appears in another specification dated 1794.[146] Amongst writers on artists' colours, de Massoul was the first to mention Mars yellow, but, although his book usually supplies some kind of explanation of how a pigment is made, in this instance it contains only the brief comment that it is made from 'a dissolution of iron, and afterwards precipitated'. In de Massoul's opinion the process was known to perfection by only a few people, including himself, and he intended to give nothing away. In 1801 Ackermann was selling Mars yellow water-colour cakes, but, apart from Varley's mention of iron yellow in 1816, there is very little concerning the pigment to be found in English sources. There is no mention of the colour in Field's *Chromatography* and two examples in his 'Practical Journal 1809', ff. 361 and 377, are described as French, so Hurst's statement that Field introduced a series of Mars colours must be true for a later period.[147] French Mars colours are listed in Winsor & Newton's earliest catalogue; therefore it seems likely that until about 1835 the yellow, which de Massoul describes as light, transparent and easy to use, was a French speciality.

Gold

Genuine gold was sometimes employed as a pigment during the seventeenth century, although less frequently than at a slightly earlier period. Sixteenth-century manuscript sources and seventeenth-century copies contain numerous instructions for writing gold letters and to a lesser extent for laying gold leaf. At a time when limning was associated with the goldsmiths' craft and portrait miniatures were set in jewelled frames, a quantity of genuine gold paint was in keeping, but after Hilliard's time gold was mentioned far less and there was obviously a feeling that it was not acceptable in water-colour painting. It is true that Norgate used it sometimes and he recommended that it should be ground with gallstone to represent gilded armour, but general opinion against the use of water-colour gold is reflected in Goeree's comment:[148]

> Shell-Gold, because there is no colour that surpasses gold, either in virtue or lustre, so likewise you shall only make use of it in its extremest heightnings, or lights (imitating such rules as hereafter shall be dictated unto you) and in these you shall be sparing of your gold also as much as possible, for many with using too much gold in heightning up their work have spoiled it utterly; therefore my Council to them that will exercise themselves in Painting with water-Colours is, to banish gold quite from all their work.

Orpiment was the bright yellow pigment generally recommended for imitating gold, although on rare occasions massicot, a lead-based yellow, would be suggested.

Orpiment and King's Yellow: As_2S_3

Referring to orpiment, Peacham mentions the Latin names *arsenicum* and *auripigmentum,* explaining that the latter describes the mineral in its natural state 'because being broken it resembleth Gold for shining and colour'. Natural deposits of orpiment (yellow sulphide of arsenic) occur in various places in Europe and Asia, and the pigment has been used since early times. Nevertheless, it was the most unpopular of all traditional colours because of its poisonous nature, for, whereas other poisonous colours were used without much complaint, orpiment was thoroughly disliked because it was always accompanied by dangerous fumes and an offensive smell. The ill effects of the arsenic which it contained were well known, and, although physicians made use of arsenic as a drug in carefully measured doses, many artists preferred to avoid all contact with it.

An additional defect was the incompatibility of orpiment with a number of other colours containing copper and lead, and numerous warnings against such mixtures are to be found throughout literature on painting. The following example is taken from one of the early manuscript sources:[149]

> In Limning you must grynde your orpiment with gum water or of arrabick that is fine . . . and it is the best cleane culler of it self with out any other and it may not lye uppon Verdigrece greene nor uppon bise nor uppon red lead nor uppon russett for theis will sley the Culler of the orpiment.

It is noticeable that it was always the other colours that were blamed for the trouble.

Orpiment is frequently included in colour lists or instructions for heraldic blazonry in books, possibly because it could be used in isolation for that type of painting, but it was not recommended for portrait miniatures. Both Hilliard and Norgate state quite definitely that it should be avoided. It was used in oil painting, often for drapery, either alone or in conjunction with yellow ochre, but it was a poor drier and required some addition to improve it. According to de Mayerne, Cornelius Johnson recommended that orpiment should be ground in oil previously boiled with litharge and Van Dyck recommended the addition of the traditional (but now discredited) artificial drier, ground glass, which is also mentioned by Haydocke.[150] Bardwell refers to king's yellow in his discussion of yellow and green drapery in *The Practice of Painting,* and orpiment has been identified in some of his portraits where it has been used for representing highlights on metallic objects.[151]

Instructions for making sulphide of arsenic are given in detail by Dossie and de Massoul, both of whom call it king's yellow as was customary during the eighteenth century. References to artificial orpiment and the name 'king's yellow' have not been found in English language sources prior to the eighteenth century, which is extremely surprising in view of the fact that the orpiment which Cennino knew in the fourteenth century was manufactured and that the Dutch equivalent of the name 'king's yellow' was used in the seventeenth century.[152] The name is obviously associated with the symbolism of alchemy, current then and many centuries earlier, for Arab alchemists of the Middle Ages applied the phrase 'the two kings' to orpiment and realgar.[153] The only explanation for the apparent absence of the manufactured pigment from English sources is that until the eighteenth century the British called both natural and manufactured yellow sulphide of arsenic by one name: orpiment.

Documentary evidence dating from the eighteenth century suggests that the manufactured pigment might have been used more then than the native mineral, although quantities were still imported—most from Holland, some from Germany and small amounts from Italy or Venice. During the early nineteenth century, good supplies of the mineral were imported from China. Sold as Chinese yellow, it was extravagantly praised in Roberts' *Introductory Lessons* as the 'most beautiful colour we have'. The best orpiment was an extremely brilliant yellow, but its defects made it generally undesirable, and virtually all the new yellows introduced during the eighteenth and early nineteenth centuries were recommended for their ability to replace orpiment and king's yellow.

Turbith Mineral: Hg_3SO_6

The name turbith mineral probably developed from the seventeenth-century use of basic sulphate of mercury as an emetic, because a vegetable product called turbith, which was prepared from the East Indian jalap tree, was also used for the same purpose. In view of the medicinal use of the mineral product, it is not surprising that de Mayerne should have been the first to mention its possibilities as a pigment. He describes it as an excellent yellow, suggests that when it fades its colour can be restored by heating, and comments on the fact that it must be well washed, without explaining, however, that it is the washing process which produces the yellow colour.[154] For manufacturing instructions one must refer to Dossie, who claims to have recommended its use as a pigment to several painters. He explains that it can be made by heating equal quantities of mercury and sulphuric acid in a retort until the mercury is reduced to a white mass, which must be powdered and washed until it turns yellow. It was obtainable from chemists' shops but not from colourmen, and the absence of any price after a description of the colour in *The Artist's Repository* suggests that, in spite of Dossie's recommendation, it was never in sufficient demand to be stocked by colour shops. Turbith mineral is occasionally mentioned in literary sources of the late eighteenth century, but, since their information was taken from Dossie's book they contain no additional details concerning the pigment. Field's 'Practical Journal 1809' contains a weak lemon, rather faded water-colour wash on folio 359, but there is really no reason to regard turbith mineral as anything other than an experimental colour of the seventeenth and eighteenth centuries.

Lead-Based Yellows

A variety of yellow compounds of lead have been used as pigments, including lead stannate (lead-tin yellow), lead monoxide, lead antimonate (Naples yellow) and, at a later date, lead oxychloride (patent yellow) and lead chromate (chrome yellow). It is difficult to discuss the first three owing to unclear nomenclature in English and other languages, and anyone studying old books on painting either in the original or in translation must be aware of the problems.

Of the three pigments, lead-tin yellow (Pb_2SnO_4), lead monoxide (PbO) and Naples yellow ($Pb_3(SbO_4)_2$), nineteenth-century writers and translators were perhaps most familiar with the last, and this led some to translate the Italian colour name *giallolino* as Naples yellow, as Lady Herringham did in her translation of Cennino Cennini's *Il libro dell'arte.* Mrs Merrifield, who did so much valuable work in the study of treatises on painting, came to the conclusion that there were three kinds of *giallolino:* lead antimonate (Naples yellow), lead monoxide (massicot) and a variety of lead stannate she called vitreous yellow that she had encountered in the Bolognese MS. 273. She wisely concluded that it is scarcely possible to decide on the identity of *giallolino* with any certainty when the colour name is used without further description.[155]

Still in the nineteenth century, Eastlake touched on a similar problem in regard to the name *massicot* in several north-European versions as used by writers such as van Mander and Hoogstraten, expressing surprise that they did not refer to any pigment that might be intended to represent Naples yellow, thus implying that he interpreted *massicot* as lead monoxide.[156]

In the first half of the twentieth century, Thompson sensibly left the name *giallorino* in the original in the text of his English translation of

Cennino's work but added a note that the pigment might be interpreted as *massicot,* that is, lead monoxide.[157]

During the second half of the twentieth century, a tendency has developed to interpret the names *giallolino* and *massicot* without question as lead-tin yellow as a result of publication in 1941 (and subsequently) of the identification of this pigment in paintings.[158] While it is true that during the second half of the twentieth century those engaged in the analysis of pigments have found many notable occurrences of lead-tin yellow in north- and south-European paintings dating from the fourteenth to the eighteenth century, and it is incontestable that lead-tin yellow was extensively used during that period even though little documentary evidence on the subject has been found, it must be borne in mind that lead monoxide has occasionally been identified in paintings and that lead antimonate has also been identified in paintings from the seventeenth century onwards, and, likewise, their use is not very well supported by documentary sources.[159]

Original treatises indicate quite clearly that there were several types of *giallolino* and *massicot.* Such is the twentieth-century preoccupation with the relatively recently rediscovered lead-tin yellow that at least one writer has interpreted this as reference to two varieties of lead-tin yellow: that prepared from metallic lead and tin and that prepared from molten yellow glass (Kühn's lead-tin yellows I and II).[160] However, this does not really seem to be tenable, as the vast majority of notable occurrences of lead-tin yellow have been identified as the former, type I. It would seem much more likely that the admission in a considerable number of original treatises that several varieties of *massicot* and *giallolino* existed serves to indicate that more than one pigment went by the same colour name and that reference may be to any one of the three pigments: lead stannate, lead monoxide or lead antimonate. The first named, that is the one commonly known as lead-tin yellow, may be the most likely interpretation, but, in a case where the English colour name *massicot* is used alone without qualification, there can be only one certainty, and that is that a lead-based yellow is indicated.

Massicot and General

The name *massicot* appears frequently in English books of the seventeenth century, but those written *circa* 1600 or earlier also contain another name—*general*—which was applied to the same pigment, for Peacham speaks of 'masticot or general', obviously treating the names as synonyms. Both names are mentioned by Haydocke, who states that massicot and general were made in Flanders and Germany, and both are listed in V. & A. MS. 86.EE.69 where *masticotte* is quoted at six shillings and eightpence per pound and *generall* is listed at half the price for the same quantity. If massicot and general were identical, the price difference is surprising; it supports the view that the names may have been attached to different types of lead-based yellow.

English books supply no instructions for making the pigment and only

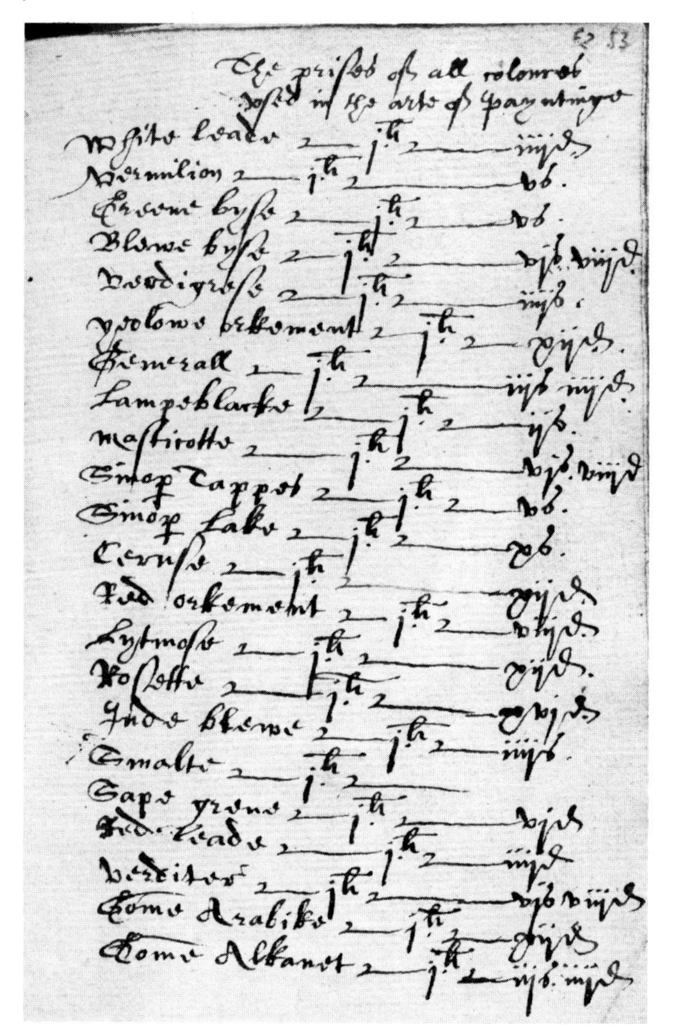

The prises of all colours
used in the arte of payntinge

White leade — iiijd.
Vermilion — vs.
Greene byse — vs.
Blewe byse
Verdigrese — iiijs.
yellowe orkement
Smeralt
Lampeblacke
Masticotte — ijs.
Sinoper Lappet
Sinoper lake — vs.
Ceruse
Red orkement
Lytmose
Rosette
Inde blewe
Smalte — iiijs.
Sape greene
Red leade
Verditer
Gome Arabike
Gome Albanet

Fig. 29. Late sixteenth-century colour list with prices ranging from fourpence per pound for white and red lead to ten shillings per pound for the red, sinoper lake. The list concludes on another page, not illustrated, with dragon's blood at six shillings and eightpence per pound and gold and silver foil (Victoria & Albert Museum MS. 86.EE.69, f.62). (Reproduced by courtesy of the Victoria & Albert Museum.)

one reference gives an indication that it contained tin. Waller's paper in *Philosophical Transactions* includes the statement 'Massicot is an improper calx of tin' and in the Latin version supplies a reference to a work by Van Helmont, a seventeenth-century physician. The relevant passage appears in one of his works which had been translated into English.[161]

> For just even as Tinne (which affords to Painters a yellow colour, which they call Masticot) makes every mettall (its lead being taken away) brickle: So also it tingeth the hardnesse of Gemmes, or precious Stones.

Lead-tin yellow could be prepared in different degrees, varying from a lemon yellow to a much deeper yellow. This may account for the fact that many seventeenth-century writers refer to *masticots* in the plural. Documentary sources supply some conflicting information on the subject of grinding; most authorities consider that the colour should be ground very little or not at all for use in oil or water colour, whereas Marshall Smith states that massicots are not fine enough as sold in the shops and that the oil colour must be ground more finely in linseed oil and then stored in bladders. There was no disagreement about the quick-drying property of massicot, and a number of writers point out that if it is not kept in bladders it must be placed under water to prevent oxidation. Because the colour was obtainable in different degrees, seventeenth-century painters found it useful for a number of purposes, but during the following century it was criticised for lack of brilliance and as a consequence it was used less then than at an earlier period.

Documents on water-colour painting also provide a number of conflicting opinions. Hilliard states that massicots of different degrees are the best yellows; they should be washed, ground and used with an addition of sugar candy. Norgate claims that the pigment should not be ground because, if it is, the paint is liable to become 'of a greasy and clammy thickness'. After the pigment is washed, it should be tempered with gum.[162] Goeree states that massicot is not much use in limning, and *A Book of Drawing,* 1666, contains the information that light massicot is not an essential colour for washing prints. Massicot suffered from the same defect as lead white in water colour in that it tended to become grey and discoloured on exposure to air; a good example of darkened massicot is to be seen in de Mayerne's manuscript, B.M. MS. Sloane 2052 (*Plate 2*). Nevertheless, despite this deficiency, a lead-based yellow of this name was in use throughout the seventeenth and eighteenth centuries. By the end of the eighteenth century it seems reasonable to interpret massicot as being lead monoxide, as so few occurrences of lead-tin yellow have been encountered in the late eighteenth century that it must have been virtually obsolete, and lead antimonate was definitely called Naples yellow by that time. Ackermann supplied massicot in water-colour cakes at the very beginning of the nineteenth century and Field's 'Practical Journal 1809', f. 369, contains a sample of colour that was obtained from the artist James Ward and said to be highly prized by him, but one must assume that it became obsolete shortly afterwards as it does not appear in colourmen's lists of oil or water colours in the 1830s.

Naples Yellow: $Pb_3(SbO_4)_2$

English sources of the seventeenth century contribute nothing to the knowledge of Naples yellow, a colour which was the subject of controversy in the eighteenth century when it was variously supposed to be a type of ochre, a sulphurous product of Vesuvius, or a manufactured colour. Although the pigment was known on the Continent in the seventeenth century and it is mentioned in the books by Boutet and Goeree, it is not listed in any book by an English writer during that period. This raises the question as to whether it was perhaps known in England but loosely referred to as one of the several types of massicot; writers on painting may not have felt a need to distinguish between one lead-based yellow and another. Certainly during the eighteenth century,

English writers appear to have had differing ideas concerning the pigment, as Elsum states that it is a middle colour between massicot and light ochre and that it is most suitable for miniature painting, whereas Dossie states that it is used mainly in oil. The composition of Naples yellow was investigated on the Continent during the second half of the eighteenth century, and all the evidence to be found in English sources is second-hand. An Italian, Passeri, was for long reputed to have been the first to establish the fact that the pigment was manufactured and that it contained lead and antimony, although his recipes are based on those of Piccolpasso (written 1556–1559), or a source very close to him.[163] In 1766 de Bondaroy in France published a paper that contains numerous references to traditional views on the composition of Naples yellow and also gives details of his own experiments and method of making the colour.[164] From that time onwards, Naples yellow was generally recognised as lead antimonate, a manufactured colour, but only occasional references to it are to be found in English literary sources. The tradition that it would change colour if it came in contact with iron is to be found as far back as Dossie's *The Handmaid to the Arts,* and Payne remarks that the pigment 'of a very sickly hue' is gritty and hard to grind and prepare. Field's 'Practical Journal 1809', f. 347, contains an example of the colour that was obtained from the artist Reinagle.

Patent Yellow: $PbCl_2 \cdot 5\text{–}7PbO$

A few years before his discovery of a new green, Scheele discovered a process for preparing soda whereby a yellow pigment, lead oxychloride, was produced as a by-product. The discovery was made about 1770 and was made public some five years later.[165] After a similar interval, the process was patented in England by James Turner, actually in February 1781, not 1780 as stated by some authorities.[166] It was described as 'a method of producing a yellow colour for painting in oyl or water, making white lead, and of separating the mineral alkali from common salt, all to be performed in one single process'. A quantity of lead oxide (both red lead and litharge were mentioned) and half its weight of sea salt in solution were ground together and were allowed to stand for twenty-four hours, after which the caustic soda solution was poured off, and the remaining white substance was calcined until it reached the desired shade of yellow.

Turner benefited very little from his patent at first, because the process was copied by other manufacturers. He brought and won an action against one of his competitors, but, following an appeal against the verdict in 1787, another trial was ordered. The appeal has received attention in books on patent law, because it was treated as a ruling case during the nineteenth century, the rule being that if a patentee claimed to do several things by one process and one failed then the whole patent was void.[167] Turner had listed several alternative names for lead and also for the type of salt which could be used, and the plaintiff held that the specification was insufficient because some types of the lead oxide and salt mentioned would not produce the pigment according to Turner's instructions. The appeal was upheld and a new trial ordered, but Turner finally won his case and was allowed to retain his patent. He obtained an extension of the time he was allowed sole right of manufacture on the ground that he had not benefited from his patent because certain 'chemists and colourmen' who were in possession of the channels of the trade had manufactured the pigment with the result that over a period of years Turner's income had not covered his expenses. The extension was obtained by Act of Parliament, the only means by which a patent could be prolonged. In the statute Turner's yellow is described as both better and cheaper than orpiment, as being made entirely from British raw materials and as contributing to the national income through export and by contribution to the salt tax.[168] Accounts of the legal proceedings give no indication of the place where Turner's yellow was made, although, according to one authority, the pigment was manufactured at the soda factory run by Losh at Walker-upon-Tyne where it was sold as Turner's Patent Yellow.[169]

It was not long before the pigment was known in England, for it is described in Williams' book on oil painting as 'a new invented colour, very bright and durable; dries well'. Its lead content would have contributed to its good drying qualities in an oil medium, but the lead was also responsible for a tendency to blacken on exposure when used in water colour. Nevertheless, it was supplied in that form, as it was advertised by Ackermann at one shilling per cake. Field commented that patent yellow had good body and worked well in both oil and water colour but was soon injured by sunlight and impure air, and was, therefore, little used 'except for the common purposes of house-painting'. In spite of the introduction of other yellows in the nineteenth century, patent yellow was still made on a large scale. In the middle of the century it was manufactured at Washington, County Durham, by the Washington Chemical Company, which displayed specimens of the colour together with illustrations of its use in oil painting at the Great Exhibition of 1851.[170] Patent yellow oil colour appears in colourmen's catalogues dating from the first half of the nineteenth century, but lack of references to it in literary sources of the same period suggest that it was to some extent superseded by chrome yellow.

Chrome Yellow: $PbCrO_4$

A new yellow was discovered when chemists were investigating the properties of a mineral, a natural lead chromate now called chrocoite, which was found in the Beresof gold mine in Siberia in 1770. Pieces of the mineral, which varies from orange-yellow to orange-red, were occasionally used in painting, although, as the mine was not working continuously, there was no regular supply. Various chemists analysed the mineral with different results, but, in 1797, at his second attempt,

Fig. 30. Louis Nicolas Vauquelin (1763–1829). (Reproduced by courtesy of the Ann Ronan Science Library.)

Vauquelin discovered that the substance contained an unknown element. It was isolated by him the following year, and it was called *chrôme* on account of the different coloured precipitates (green, red and yellow) which could be obtained from it.[171] Lacking supplies of the ore, Vauquelin was unable to pursue his research as he wished, but by 1809 a source had been found in the Var region of France and he was able to continue his experiments. He gave a full description of the preparation of the yellow pigment, lead chromate, which was precipitated when lead acetate or nitrate was added to potassium chromate, explaining that slight adjustments of the process would vary the tone of the colour. Vauquelin's paper on the subject contains the statement that there was no necessity to discuss the different varieties of lead chromate used in painting at that time because their beauty, good working quality and permanence were then quite well known by painters.[172]

In spite of Vauquelin's remark, there is general agreement among later writers that chrome yellow was not commercially available immediately after its discovery owing to a scarcity of the raw material. Nevertheless, some writers have suggested that it was first available somewhat later than was actually the case, the general impression being that it was not available until quantities of chrome ore were discovered in North America as late as 1820.[173] Evidence in Field's 'Practical Journal 1809' shows that by 1814–1815 there were several sources of supply of chrome yellow and that the British artists Thomas Lawrence and William Beechey and the Americans Benjamin West and Washington Allston all had some at the time. One of Lawrence's samples called Paris yellow, came from France, while West had some chrome ore from Pennsylvania that Humphry Davy used to make some of the yellow pigment for him, and Allston's sample was made up by Bollmann in England.[174]

Dr Bollman (1769–1821) is accepted as being the first person to manufacture chrome yellow commercially in England. His career is of some interest, not merely on account of the new pigment. He was born in Germany and obtained the degree of doctor of medicine at Göttingen, after which he spent some time travelling in France and England. As a young man inspired by current revolutionary ideals, he played an important part in arranging the escape of the French liberal leader, Lafayette, from Austrian imprisonment at Olmütz, which, according to some French authorities, is Bollmann's real claim to fame. The incident had tremendous influence in shaping Bollmann's subsequent career, because, becoming *persona non grata* in Germany as a result, he decided to emigrate to America, the United States being the obvious choice for anyone of liberal political opinions. He arrived there in 1796 and remained until the European war ceased in 1814, when he returned, spent a short time in Vienna and then settled in London. He was doubtless at home in an English-speaking country, although his correspondence was in German, but in any case, he regarded England as a land of opportunity, as he wrote in 1816: 'im chemischen Manufaktursach ist hier noch Vieles zu thun' (there is still much to do here in chemical industry). Bollmann wrote to his brother, who was in Philadelphia, and to others, and his letters give an insight into his manufacturing activities in London. Bollmann's American sojourn and connections were important in that they enabled him to bring a quantity of chrome ore to England and to set up a private laboratory where a workman under his guidance daily produced twenty pounds of lead chromate, 'die schöne, neue, gelbe Farbe' (the fine, new, yellow colour) as Bollmann described it. His livelihood did not depend on his laboratory, for he entered into a business association with a rich young Englishman, who was so interested in chemistry that he financed the operation, and a Swabian who had been trained in chemical manufacture in France. They set up a factory on the Thames at Battersea where pyroxilic acid, verdigris, lead acetate, soda and other products were made. Although Bollmann claimed that he and his associates made a profit in the region of 100 to 200 per cent as a result of their superior methods, the running of a large factory was attended by difficulties, including local opposition concerning air pollution and antagonism to the workforce, which was entirely German.[175] So Bollmann was able to devote only part of his time to the private laboratory; nevertheless, his private enterprise was successful, and Field testifies to the importance of Bollmann's contribution to chrome yellow manufacture, saying that not only did he introduce the pigment but that he collaborated with Field in experiments to improve the colour. It is not known if Bollmann continued to obtain supplies of chromium from the United States, but, in any case, the discovery of chrome ore in one of the Shetland Islands meant that there were ample supplies in the British Isles after 1820 and Berger too was manufacturing the pigment by 1824.[176]

Chrome yellow can vary from a light yellow to a strong orange-yellow, and the latter variety was especially welcome as an alternative to orpiment and also to patent yellow. The fact that chrome yellow was poisonous, although to a lesser degree than orpiment, did not cause any concern, the main source of disappointment being the tendency of the pigment to become discoloured on exposure. The sample of chrome yellow in *Le maître de miniature,* 1820, shows signs of deterioration in patches. The colour is mentioned by Fielding, who lists it amongst the thirty most useful colours for landscape and figure painting; the colour sample in his book is in good condition. In 1835 it was listed with the cheapest water colours at one shilling per cake, and its low cost has doubtless contributed to its continued use as an artists' colour even though some subsequently discovered yellow pigments are more permanent.

Platina Yellow: K_2PtCl_6

In the eighteenth century platinum was obtained from alluvial deposits in one of the Spanish American colonies, now Colombia, and an account of the metal was published in 1750.[177] During the nineteenth century platinum was used to simulate silver plate on pottery, and it is possible that its availability as an artists' colour was connected with its use in an allied industry. Field refers to platina yellow in *Chromatography* but offers no information about its composition. Bachhoffner, who states that Field introduced the pigment in England, gives an account of its composition which suggests that it was potassium chloroplatinate, a yellow crystalline precipitate.[178] Bachhoffner also says that, at one time, Field sold it under the name lemon yellow, which he later transferred to a chromate. Subsequently, colourmen listed the colour under the name platina yellow, but, presumably owing to the cost of platinum and the ready availability of barium chromate, platina yellow became obsolete during the first part of the nineteenth century.

Lemon Yellow: $BaCrO_4$

Earlier writers occasionally likened a pigment to a lemon colour, but lemon yellow was first used as a colour name by Field, who mentions it almost in passing and gives no explanation of its composition, although there is a sample of the colour in *Chromatography*. Today, both strontium chromate ($SrCrO_4$) and barium chromate ($BaCrO_4$) are sold under the name, and, theoretically, the pigment could have been made from either during the nineteenth century, as strontium first came to the attention of chemists in the early 1790s and barium chromate was mentioned, although without reference to its use as a pigment, by Vauquelin in 1809. It seems that neither was employed in England until, according to Bachhoffner, Field chose barium chromate as a cheaper alternative to platina yellow, which he had previously sold under the name of lemon yellow.

Cadmium Yellow: CdS

A colour which could have been made before 1835 but does not appear in any colour lists is cadmium yellow. Stromeyer discovered cadmium in 1817 and discussed cadmium sulphide, which he likened to orpiment but recommended as being more durable. A number of people recognised the potentialities of the pigment, but, as Bachhoffner pointed out, scarcity of the metal postponed its manufacture and it was not commercially available until the 1840s.[179] A cadmium yellow sample is included near the end of Field's 'Practical Journal 1809', f. 385, with the comments 'Works well—acts slightly on knife—Not more changed than the paper by suspending in damp and foul air more than five months.' The entry is undated but it is a late entry, as f. 385v bears the date 1844.

8

Organic Yellows

Saffron

The deep yellow colour made from the dried stigmas of the autumnal *Crocus sativus* was one of the colours traditionally employed in manuscript illumination until the sixteenth century, when judging from instructions given in *Limming,* 1573, and in B.M. MS. Sloane 3292, it was frequently tempered with glair:[180]

> Roche Yellowe. Take fine Saffron and lay it in glaire the space of a day and a halfe, it makes a puer yellow water. Lay it uppon cleane paper or parchment, or any other culler that is whyte but not elce.

Because of its transparency, saffron was recommended for colouring prints and maps during the seventeenth and eighteenth centuries, but English water-colour painters would seldom use it for any other purpose. Hilliard admitted it as a shading yellow only if nothing better was available, and Norgate relegated it to the list of organic colours which he considered unsuitable for miniature painting owing to their fugitive nature.

Saffron yellow is mentioned occasionally in late eighteenth-century books, but at that time and during the early nineteenth century the name *saffron* was sometimes misapplied to a red made from safflower.

Fustic

The yellow dye generally known as old fustic, which is obtained from the plant *Chlorophora tinctoria,* was introduced to Europe in the sixteenth century when it was imported from the West Indies and Central America. The dye, which must not be confused with 'young' or Zante fustic which comes from *Rhus cotinus,* was employed in textile dyeing and to a very limited extent in water-colour painting. The colour is referred to as *ffusticke yealowe* under the heading 'fine yellows' in B.M. MS. Stowe 680, f. 133v.

Fustic and saffron were little used in water-colour painting and not at all in oil painting during the seventeenth century; the organic yellow known as pink was by far the most important organic yellow in both oils and water colours. However, old fustic was one of the dyes sometimes employed by Berger in the manufacture of yellow lake pigments in the early nineteenth century.[181]

Gallstone

A gallstone from an ox sometimes served as a colour for limning, when, crushed and ground in gum water, it provided a fairly dark yellow which Hilliard describes as a useful colour for shading. In Norgate's treatise it is not included in the main list of colours but it is mentioned in connection with the painting of gilded armour in portrait miniatures.[182]

> There is a Stone growing in an Ox gaule, which they call a Galle Stone, which ground and tempered with Gould is excellent for all Gould workes, and gives an excellent Luster and bewty in the Shadowing within the deepest and darkest Shadowes must bee mixt with Black.

Gallstone is mentioned in books which were copied from Norgate's treatise, but not in other books dating from the second half of the seventeenth century. At that period, the French writer Boutet referred to *pier de fiel* as a new colour.

The pigment reappears in the eighteenth-century books, but, according to Dossie, only a few painters in water colours knew of it. He recommends that if gallstones are unobtainable a similar colour may be made by boiling a quart of ox bile in a double saucepan, adding a quarter of an ounce of gum arabic and allowing the moisture to evaporate so that the dry colour is left. The recipe for the substitute, which was said to be more transparent than genuine gallstone, was taken up by colourmen, and at the end of the eighteenth century they were selling the substitute under the name gallstone. John Payne complains of this in his book on miniature painting and recommends painters to make application at the Victualling Office or any private slaughter-house for the slaughter-men to look for gallstones. Its rarity is underlined by its inclusion in Ackermann's list of 1801 at a price of five shillings per cake, that is, as one of the four most expensive colours. That reference is the last which almost certainly indicates genuine gallstone, as Varley mentions that colourmen have managed to render the colour more permanent, a remark which suggests that it contained something other than gallstone. It was possible to match the colour of gallstone with quercitron yellow, a lake pigment which was sometimes sold under the name of gallstone during the nineteenth century.[183]

Gamboge

The gum resin from varieties of *Garcinia,* evergreen trees which grow in south-east Asia, can be used as a drug and a pigment. The name gamboge, by which it is generally known, is derived from Camboja, the old name for Cambodia where the trees are found. Other names were sometimes used during the seventeenth century; in Parkinson's *Theatrum Botanicum,* in addition to the description *Cambugio quibusdam carthartícum aureum,* the golden yellow Indian purger, the alternative name *gama gitta* is supplied. This form is related to the modern French *gomme goutte* and is derived from the Malay word *gatah* meaning gum, although it has since been associated with the idea of drops because the gum resin oozes from incisions made in the bark of the tree. All early writers who mention gamboge were aware of its being a gum, and they often altered its spelling accordingly; an example is *gum-booge,* which appears in *The Excellency of the Pen and Pencil.* The last syllable illustrates the correct pronunciation of gamboge.

It is possible that small quantities of gamboge were used in England before the seventeenth century, because it could have been transported by the overland route, but it was regarded as a novelty in England when the East India Company imported a quantity in 1615.[184] The correspondence which passed between the East India Company and its factors contains references to gamboge as an article of commerce but does not contain a description of the manner in which the gum resin was

Fig. 31. Pipe gamboge. The gum resin is poured into bamboo shoots, allowed to dry and then removed in cylindrical pieces, the form in which it is exported from the East. (Reproduced by the kind permission of Messrs Winsor and Newton Limited.)

collected. Parkinson describes its appearance when it arrived in England as a 'solid peece of substance, made up into wreathes or roules, yellow both within and without'.[185] This is a good description of pipe gamboge, that is, gum resin which has been allowed to run into a hollow pipe or shoot of bamboo and has been removed from the bamboo after it has set hard.

The pigment is not mentioned by Hilliard nor by Norgate, but it is listed as an oil colour in the second edition of Bate's *Mysteryes* and it is recommended as a water colour by de Mayerne. It is frequently mentioned as a transparent yellow for colouring prints in books written during the second half of the seventeenth century. Smith describes how easily it may be prepared for water-colour painting:[186]

> For a Yellow Gumboge is the best, it is sold at [the] Drugist in Lumps, and the way to make it fit for Use, is to make a little hole with a Knife in the Lump, and put into the hole some Water, stir it well with a Pencil till the Water be either a faint or a deeper Yellow, as your occasion requires, then pour it into a Gally-Pot, and temper up more, till you have enough for your purpose.

Purely because it was easy to prepare and it required no additional gum, pieces of gamboge are sometimes to be found in addition to cakes of other colours in old water-colour boxes. In the late eighteenth century, pieces of gamboge could be obtained from a chemist at sixpence or eightpence per ounce. In painting it was used almost exclusively as a water colour, although burnt gamboge is mentioned in *A Compendium of Colors,* 1797, as a brown oil colour useful for glazing.

In the early nineteenth century gamboge water colour was mentioned by Dayes, Henderson, Dagley, Varley and a number of others. It was used alone as a yellow, mixed with Prussian blue for green, and mixed with burnt sienna for orange. The authors differ in opinion over the question of permanence, which is understandable as the durability of gamboge tends to vary from one piece to another. There are sample washes of the colour in a number of books, including those by Clark, Roberts, Varley and Fielding.

Pink

The English word *pink* was once used as a noun with reference to a yellow pigment. Only during comparatively recent times has the word been accepted as an adjective meaning rose pink or light red. Many seventeenth-century references to pink suggest that it was made from unripe buckthorn berries (of the genus *Rhamnus*), but there are a few indications that it was sometimes made from other raw materials. A comment in one of de Mayerne's manuscripts suggests that it could be made from weld.[187] Norgate had a recipe for making it from *Genestella tinctoria,* a variety of broom, and additional recipes in one of the copies of Norgate's treatise include the remark that 'callsind eg shels and whitt Roses makes rare pinck that never starves'.[188] From these alternatives it is clear that the word *pink* did not represent the raw material which provided the colouring matter. Nor did it represent a particular hue, because there are numerous examples of the use of the word with a qualifying adjective, many of which describe its hue. Yellow, green and light pink are seventeenth-century variations, whereas the names brown, rose, Dutch and English pink were current somewhat later. Brown pink survived as a colour name into the nineteenth century, when the name Italian pink was also used.[189] A common feature of all the pigments

Fig. 32. Weld plant: from Parkinson, *Theatrum Botanicum*, 1640. (Reproduced by courtesy of the Royal Horticultural Society.)

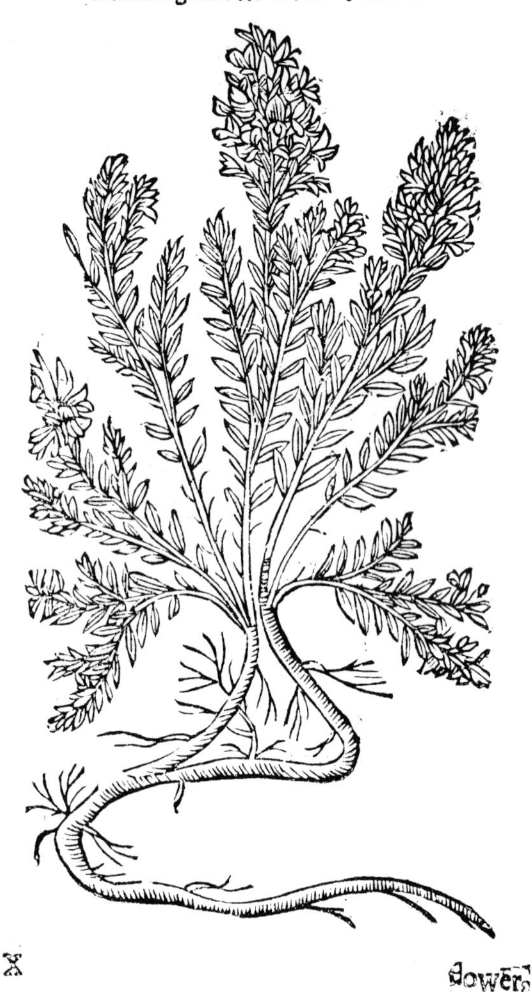

7. Genista tinctoria vulgaris.
Common greene weede, or Dyers weede.

Fig. 33. Broom plant, used in dyeing and occasionally for making the pigment, pink: from Parkinson, *Theatrum Botanicum*, 1640. (Reproduced by courtesy of the Royal Horticultural Society.)

described by those names is the fact that all were obtained from a dye that, during the early part of the period under discussion, was mordanted on chalk or alum, that is, it was not a true lake pigment.

Any discussion concerning the difference between the old and modern meaning of pink is made immeasurably more difficult by the fact that the word *lake* was once applied strictly to red or light red (modern pink) pigments. That was what it conveyed to painters, and the fact that those pigments were also what are known as lake pigments may have been incidental. Throughout the seventeenth century, lake (red pigment) and pink (yellow pigment mordanted on chalk or alum) were entirely distinct, and it was not until the early eighteenth century that the name *rose pink* was applied to a light red, pseudo-lake pigment. It appears in *The Art of Painting in Miniature* (which is a translation of Boutet's *Traité de mignature*), where it is given as the English equivalent of *laque colombine,* a light-red pigment made from brasil wood. Presumably the translator's avoidance of the word *lake*, which he used for Florentine and Venice lake, was either that it should be reserved for *lacca* of Indian origin or that it should be applied only to red pigments made by a method other than mordanting a dye on alum, which is the way the rose pink from brasil was made. In any event, the name *rose pink* was used by Dossie, also for a brasil pigment, and such was the influence of his book that the name was repeated many times during the second half of the eighteenth century. By that time the word *pink* was never used alone when referring to a pigment, but was always qualified by an adjective such as rose, brown, English or Dutch. It is possible that pink would never have come to be identified solely with light red but for the fact that during the late eighteenth century the name as applied to a yellow pigment fell into disuse and was replaced by the name yellow berries. It is true that the name brown pink remained for over a century and that other variations are to be found in specialised literature of a later date, but the passing of the popular conception of pink as yellow can be seen in different editions of *Bowles's Art of Painting in Water-Colours.* In the edition of 1783 English and Dutch Pink are listed and said to be made from French berries, but in the edition of 1786 French berries is given as the colour name.

Although Hilliard states that pink should be avoided, the colour was popular throughout the seventeenth and eighteenth centuries. All seventeenth-century sources suggest that it was a yellow which tended to green and that its main use was to provide a green when mixed with blue. Norgate insists that it is necessary to obtain the best because it is an important colour, and suggests that it should be mixed with blue bice or verditer. A mixed green of this type was so usual that bice and pink are frequently listed together with the list of green pigments in the many later copies of Norgate's treatise. This explains Browne's use of the phrase *green pink,* because his colour list was originally copied from Norgate's work; he does not mean a green lake but yellow lake mixed with blue. The same association of pink and a blue in a list of greens occurs in B.M. MS. Sloane 6284, where indigo is suggested for the mixture. The same recommendation is also made in *A Book of Drawing,* 1666.

Peacham gives instructions for making a yellow pigment from Venice berries, presumably from yellow berries imported from the Near East through Venice. Norgate states that the best pink is difficult to obtain, although he does not explain which is the best variety. There is no recipe for making the pigment in the early version of Norgate's treatise, but there is a recipe, said to have been communicated to him by the amateur artist Sir Nathaniel Bacon, in the later version. It is one of the most detailed accounts of the method of making pink in the seventeenth century:[190]

Fig. 34. Sir Nathaniel Bacon (1583?–1627): self-portrait in oil by one of the most distinguished English amateur artists of the seventeenth century (Gorhambury collection). (Reproduced by permission of the Earl of Verulam.)

About midsomer take as much greene weed called in Latin *Genestella tintoria,* as wilbe well boyled and covered in a pale of water, but let the water seeth well and be scumed before you put it in. You will know that it is well sodde when the leaves and barke will strippe from the stalke drawne through your fingers. Then take it from the fire, and powre it into a woodden bowle or pale through a cloth, till all the water be strained through, then cast the wood away.
Take this water and set it on the fire againe, and when it begins to seeth, put into it the quantity of halfe an egshell of grounde chalke, or else the powder of egshell finely ground. Mingle this chalke with a little of the water of your kettle in a dish after the manner of thickning the pot, then put to it a little ielleyd size, broken small with your hand and as it were strewed all over the superficies of your colour, and soe

> let it stand. This size is put in to make the water seperate from the Colour. Then take off the scume and put it into a jarre glasse and set it where noe sun comes and it wilbe an excellent yellow.
> But the maine colour is that which sinks to the bottome, from which you must (after it hath stood shelving an howre or two) draw away the water by philter. Then powre out the rest into a thicke linnen bagge, setting a dish underneath, because the first drayning will cary colour with it, which you may after put into the bagge againe; and soe let it hang 24 howres, or twice soe long if need be. Then take it out of the bagge and slice it and lay the slices upon Cap-paper in a dish, and dry it in an Oven after the bread is drawne, and keepe it for your use. At Midsomer the herb is in Flore.

A colour made according to such instructions would not be a true lake pigment, which is always composed of aluminium hydroxide as a result of a chemical reaction between alum and an alkali such as potassium or sodium carbonate. Seventeenth-century English instructions for making yellow or red pigments from dyes seldom include an alkali but mention alum, or sometimes another base such as chalk, none of which absorbs a dye as readily as aluminium hydroxide. Many seventeenth-century instructions are so brief that one might think that the alkali had been overlooked, but in a detailed method such as that attributed to Sir Nathaniel Bacon it seems inconceivable that all mention of an alkali was omitted accidentally, and one must assume, therefore, that some pigments were produced from dyes without following a proper lake-making procedure.

All the evidence on this subject points to the fact that the pigment *pink*, in its various shades, indicated what might be called a pseudo-lake pigment. This is also borne out by the fact that Merret quite clearly understood the difference between a pigment of the type known as pink and a true lake, that is, a pigment made by a special process, when he translated Neri's book in the seventeenth century. It contained instructions for a sound method of making a lake pigment from a liquid yellow dye, and Merret translated it as 'A Yellow Lake to Paint, from Broom Flowers'. Alkali, which is absent from English recipes for pink, is listed as the first constituent in the lake-making instructions of Italian origin:[191]

> Make a Lee of Barillia, and of Lime, reasonable strong; and in this Lee, boil at a gentle fire fresh Broom Flowers, that the Lee may draw to it all the tincture of the Flowers, which you shall know by taking the Flowers out and seeing them white, & the colour well taken out, and the Lee will be yellow like good Trebian wine: then take out these Flowers, and put this Lee in earthen dishes (glased) to the fire that the Lee may boil, and put into it so much Roch-Alum, that with the fire, all the Alum may be dissolved; then make a fire, and empty this Lee into a vessel of clean water, and it will give a Yellow colour at the bottom.

Then follow instructions for washing the pigment thoroughly, changing the water several times, and for drying it, 'and you shall have a beautiful

Lake of a Yellow colour, for Painters, and also for glass'. Merret's comments on the passage which he translated from Italian include the remark that it is best to use lime and barilla (an alkali obtained by burning either the plant of the same name or kelp) although alum and potash will serve instead, so it seems that the significance of the alkali was not lost on the English. A further observation made by Merret points to the fact that lake-making procedure was known in England:[192]

> I know an Ingenuous gentleman, who this way hath made all his colours for plants, which he hath drawn to the life in a large volume of the most beautiful flours of all sorts in their proper and genuine colour.

The reference is likely to be to Alexander Marshall, the amateur painter who had refused to help the Royal Society with their proposed history of artists' colours, for he was well known as a flower painter during his lifetime. He is mentioned by Sanderson, and his book of flowers, now in the royal collection at Windsor, was known to Evelyn as well as Merret.[193]

English sources dating from the second half of the seventeenth century state, almost without exception, that the colouring matter for pink is obtained from yellow berries. Presumably the raw material had become more easily obtainable, because the berries were imported from France by that time. The yellow known as Dutch pink was considered better than English pink because the latter was paler and coarser in consistency. They were most in demand for water-colour painting, particularly for colouring prints when their great transparency was extremely useful.

In his recent research into pigments used for house painting, Bristow has been able to clarify some details regarding the varieties of pink in the late eighteenth and early nineteenth centuries as a result of his study of the records of Berger, the colour-manufacturer. At the end of the eighteenth century, Berger's English pink was made with a mixture of old fustic and quercitron dye on a base of whiting, while yellow berries were used for Dutch pink until 1815. More expensive was Italian pink that was a true yellow lake usually made from yellow berries on aluminium hydroxide. However, there are some instructions for making it from weld with the comment that this was the composition of the Spanish yellow pigment for the colourman, Newman.[194]

Brown pink was a popular colour with artists in the second half of the eighteenth century and in the nineteenth century. Eighteenth-century books are rather vague on the subject, suggesting that the same type of berries of the genus *Rhamnus* was used without explaining how a brown was produced. Berger's records are useful in this respect, as they show that first berries and, after 1819, quercitron bark were used, adsorbed on to a base of aluminium hydroxide in the presence of pearl ash. Ferrous sulphate was used as the mordant rather than alum and this would have produced the brown colour. The following are instructions *circa* 1817:[195]

> **Brown Pink**
>
> 56lb Turkey Berries and 1 Pail Second Stuff
> a Tub White

1lb Pearl Ash
1½lb Copperas
The Berries and ash put in copper and boiled 3 hours then boiled in ¾ Copper to a pap, strain and add the White and Copperas and boil down to a thick state, then put in stove in shallow pans and not put in too violent a heat or it will ferment and cause it to be more of a yellow cast.

The brown pink made by Berger for Newman in 1819 contained quercitron bark as the source of dye and 4½ oz ferrous sulphate instead of the 1½ oz used in the brown pink for regular sale.[196]

Brown pink was used extensively as a glazing colour in oil painting and the following remarks of Bardwell, which are repeated in several books on oil painting dating from the second half of the eighteenth century, probably served as a general guide:[197]

> Brown Pink is a fine glazing Colour; but of no strong Body: In the Flesh it should never join, or mix with the Lights; because this Colour and White antipathize, and mix of a warm dirty Hue; for which Reason their Joinings should be blended with a cold middle Teint. In glazing of Shadows, it should be laid before the other Colours that are to enrich it: It is one of the finishing Colours, and therefore should never

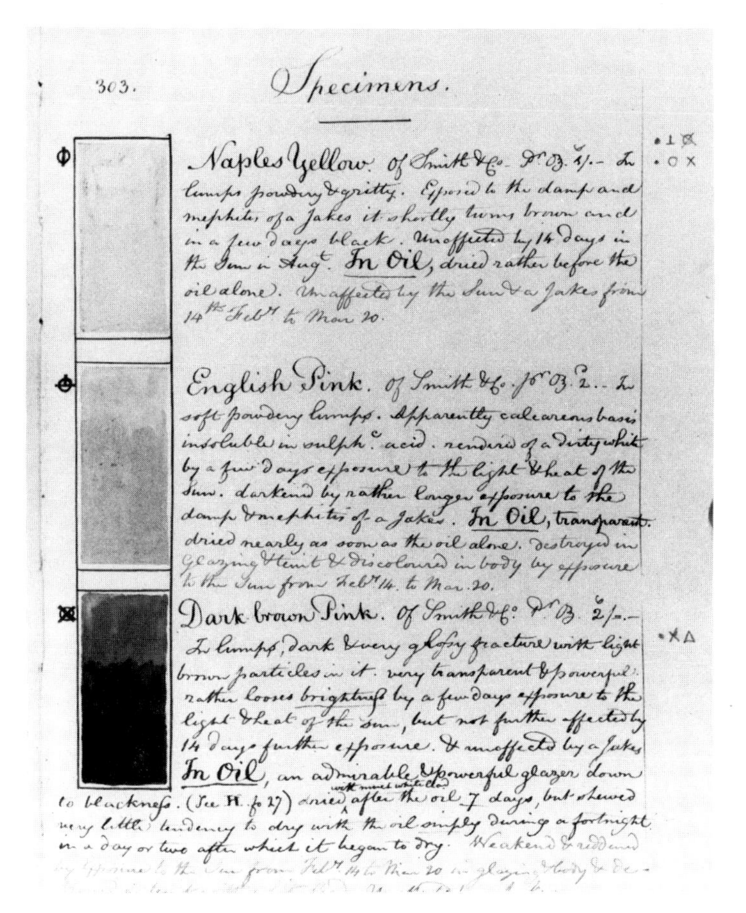
303. Specimens.

Naples Yellow. of Smith & Co. Pr Oz. 1/.– In lumps powdery & gritty. Exposed to the damp and mephites of a Jakes it shortly turns brown and in a few days black. Unaffected by 14 days in the Sun in Augt. In Oil, dried rather before the oil alone. Unaffected by the Sun & a Jakes from 14th Feby to Mar 20.

English Pink. of Smith & Co. pr Oz. 2.. In soft powdery lumps. Apparently calcareous basis insoluble in sulphc. acid. rendered of a dirty white by a few days exposure to the light & heat of the Sun. darkend by rather longer exposure to the damp & mephites of a Jakes. In Oil, transparent. dried nearly as soon as the oil alone. destroyed in glazing & tint & discoloured in body by exposure to the Sun from Feby 14. to Mar. 20.

Dark brown Pink. Of Smith & Co. Pr Oz. 2/6.– In lumps, dark & very glossy fracture with light brown particles in it. very transparent & powerful. rather looses brightness by a few days exposure to the light & heat of the sun, but not further affected by 14 days further exposure. & unaffected by a Jakes In Oil, an admirable & powerful glazer down to blackness. (See H. fo 27) dried with much white lead after the oil 7 days, but shewed very little tendency to dry with the oil simply during a fortnight in a day or two after which it began to dry. Weakend & reddend

Fig. 35. Field's comments on the poor permanence of English and brown pink in his 'Practical Journal 1809' (Courtauld Institute of Art, MS. Field/6, f.303). (Reproduced by courtesy of the Courtauld Institute of Art.)

be used in the First Painting. It is strengthened with burnt Umber, and weakened with Terraverte; ground with Linseed Oil, and used with drying Oil.

The fugitive nature of brown pink and the other varieties of yellow berries and pink is mentioned in a number of books, but it does not appear to have detracted from the popularity of the yellow in water-colour painting and the brown in oil. It is difficult to assess exactly when yellow berries or brown pink made from the berries became obsolete, because quercitron dye was used extensively in manufacturing yellow lakes during the nineteenth century and it is clear from Berger's records that the pigment then sold as brown pink was made from quercitron.

Quercitron Yellow

The brownish-yellow dye obtained from the bark of various species of American oak tree was introduced to England, and thus to Europe, by Edward Bancroft, a doctor of medicine and Fellow of the Royal Society, who was born in America but spent much of his life in England. Bancroft first tried to ship the dye to England already prepared, having extracted it from the ground, inner part of the bark. However, it deteriorated on the journey, so he decided to have the dye prepared in England, and in 1775 the first cargo of quercitron bark was shipped from Philadelphia to London. The oak from which it came was variously called yellow oak or black oak, but Bancroft named it quercitron after *quercus,* oak, and *citron,* yellow, and the name was given a form of official sanction when it appeared in letters patent granting Bancroft sole right to prepare and sell the yellow dye.

The discovery of the American product was important in the dyeing industry, because quercitron was free from any tendency to turn black if iron was added to it and it yielded a greater quantity of dye in proportion to the raw material used than any traditional dyestuff; one part of quercitron would yield the same amount of dye as ten parts of weld or four parts of old fustic. Its advantages were soon appreciated by dyers, but Bancroft's business project was unsuccessful, partly because he had omitted to obtain sole right to import the bark and partly as a result of the War of American Independence, for, soon after he was granted the patent, all trade with the American colonies was suspended. During hostilities the stock of bark collected by Bancroft's brother at Philadelphia was used as fuel by British troops. In 1785 Bancroft was once again able to import the bark, but others competed with him even though he obtained an extension of the patent by Statute in that year.[199] It has been suggested that Bancroft's extension was arranged for him in token for his services as a British agent during the American war.[200] He had acted as a double agent paid by both sides, but, according to his own account, he was always in financial difficulties and dogged by bad luck, so much so that he presented a petition for a further extension of his patent

Fig. 36. Yellow berries and quercitron bark, both raw materials for the dull yellow or yellow-brown known as pink. (Reproduced by the kind permission of Messrs Winsor and Newton Limited.)

in 1798. On that occasion he was unsuccessful, and thereafter the production of quercitron dye was open to anyone.[201]

All Bancroft wrote on the subject of quercitron, both in the patent specifications and in his book *Experimental Researches concerning the Philosophy of Permanent Colours,* describes its use in textile dyeing; there is no mention of its preparation as a lake pigment for painting. It is difficult to say when the pigment was first used, although in view of the hardships which Bancroft encountered it is unlikely to have been available commercially during the fifteen years following its introduction as a dye. Even after that time it is difficult to identify in colour lists, because the pigment was not called quercitron. The unidentified yellow lake listed in Williams' book might be quercitron, but the earliest reference with which quercitron can almost certainly be identified is the following description which appears in the treatise on Ackermann's water colours printed in 1801:

> Ackermann's Yellow, another new Colour, lately discovered, is a beautiful warm rich Yellow, almost the tint of Gall-stone, works very pleasant, and is very useful in Landscapes, Flowers, Shells, &c.

From the colour list it is clear that it was an organic pigment and the indication of its similarity to gallstone definitely suggests the appearance of quercitron. Henderson's mention of a yellow lake which can be used to glaze bright yellows in flower painting also points to a brownish-yellow, possibly quercitron. In spite of the lack of references to quercitron, the lake pigment was certainly used during the nineteenth century; it is recommended as a good yellow lake by Field, who is the only writer of the period to refer to quercitron by name. During the second half of the nineteenth century the pigment was used to a considerable extent and was sold under various names, including yellow lake, yellow carmine, yellow madder, Italian pink and brown pink.[202]

Indian Yellow

As its name suggests, this pigment was produced in India, where it is thought to have been known and used from the fifteenth to the beginning of the twentieth century. Its principal constituent is the calcium or magnesium salt of euxanthic acid. There is evidence pointing to its use in Dutch seventeenth-century painting—that is not altogether surprising in view of Dutch trade in the East Indies—but no evidence concerning the pigment has been found in English documentary sources of that period.[203]

Some English artists were definitely using Indian yellow in the late eighteenth century. In 1786 Roger Dewhurst, an amateur painter, wrote to a friend to thank him for sending some drawing paper and Indian yellow. He was unsure how to make it up into water-colour cakes, and his journal includes copies of letters on the subject which he wrote to Philips, another friend who was an expert in natural history. Apparently, Philips decided that it was an organic substance possibly made up from the urine of animals fed on turmeric and he suggested that it must be thoroughly washed as part of its preparation as a pigment.[204]

Fig. 37. Balls of Indian yellow. The upper one has been partially cut away to reveal the brighter colour inside. (Reproduced by the kind permission of Messrs Winsor and Newton Limited.)

For some time there was doubt about the nature of the raw material which was imported in roughly shaped round lumps, yellow inside but a dirty greenish-yellow outside which were accompanied by a smell of urine. Mystery about the origin of the colour probably resulted from the fact that it was made on a limited scale in only one place in India, and, although at least one person had tested the yellow and formed a fairly accurate idea of its origin, it was a long time before an accurate account of it was published. In the nineteenth century Mérimée remarked on the odour of the balls of Indian yellow but dismissed the idea that they were made from urine, whereas Field was one of the supporters of the theory that they were made from the urine of camels. It was not until the late nineteenth century that a systematic enquiry was followed by the publication of information concerning the pigment which was known as *piuri* or *peori* in India. An investigator made enquiries at Calcutta and was directed to Monghyr, where, in the suburbs of the town, he found a small group of cattle-owners who fed their cows on a special diet of mango leaves and water and collected the urine, which was a bright yellow colour. It was heated in order to precipitate the yellow matter, then strained, pressed into lumps by hand and dried. The cows were in an undernourished and emaciated condition in spite of occasional feeding with normal fodder, and production was very limited because the manufacturers were despised by other Indians of the same caste who kept dairy cattle. In 1883 the annual output was estimated at 100 to 150 hundredweights, but the investigator expressed doubts about the amount owing to the small number of cattle involved.[205] Manufacture of the pigment survived until the early twentieth century, when its production was prohibited.

Because the pigment was made in India, much was imported to England where it was prepared as an artists' colour, and, consequently, in Europe it was regarded as an English speciality. Bouvier claimed to have been one of the continental artists who used it first:[206]

> Ce jaune n'est pas très-connu; il vient d'Angleterre, et je suis un des premiers, je crois, qui l'ait fait connaître sur le continent. Il est assez cher; il se vend à Londres six schellings, c'est-à-dire à peu près six francs, l'once. Mais, au reste, il est tellement léger, et il foisonne si abondamment, qu'avec une once on en a pour l'usage de bien des années, car on en emploie peu.

Owing to its transparency Indian yellow was of most use to water-colour painters. Dewhurst was pleased to have obtained some of the colour, as he hoped it would serve as a transparent, light-fast yellow which was much needed for tinted drawings; unfortunately, the existing volume of his journal comes to an end in 1787 before he had had a chance to prepare and try the colour. Gartside's *An Essay on Light and Shade,* 1805, is the earliest printed book which contains a reference to Indian yellow. It is also mentioned by Henderson and Varley, who suggest that the colour is more light-fast than gamboge and that it is useful to make mixed greens or as a glaze over green in water-colour painting. Field had some doubts about the permanence of Indian yellow, stating in *Chromatography* that it is 'of a beautiful pure yellow colour, and light powdery texture . . . Indian yellow resists the sun's rays with singular power in water-painting; yet in ordinary light and air, or even in a book or portfolio the beauty of its colour is not lasting. It is not injured by foul air, but in oil it is exceedingly fugitive, both alone and in tint.' Nevertheless, Fielding lists it as one of the thirty most useful colours in landscape and figure painting and includes a sample wash.

Miscellaneous Organic Yellows

During the eighteenth century a number of yellow dyes were recommended for use in colouring prints. They include dye extracted from the roots of berberis bushes and mulberry trees, both of which were mentioned by Robert Boyle and recommended in *The Art of Drawing*, 1731. Also mentioned in the same book is a yellow obtained from celandine and another from roots of ginger.

Dossie discusses turmeric and zedoary as yellow washes for use in water-colour painting. Both names are associated with the tuberous roots of varieties of *Curcama* which grow in India and south-east Asia. *Turmeric* is a variation of the Latin name *terra merita*, which is also to be found in literature on painting. Both turmeric and zedoary are mentioned in books dating from the second half of the eighteenth century.

An orange colour, this time a lake pigment not merely a dye, was recommended by Dossie, who claimed to be its originator. The orange dye from annatto, the shrub *Bixa orellana*, was used in its manufacture, although Dossie was aware that it was not light-fast. It is mentioned in slightly later works, sometimes as annatto and sometimes as roucou.

During the early nineteenth century madder was used to a limited extent to make a yellow lake pigment. Sir Thomas Lawrence is recorded by Eastlake as having used Field's madder yellow, which that writer regarded as permanent, but, according to Field, the yellow made from madder root was insufficiently light-fast and he therefore ceased to make it. In subsequent years the colour sold under the name madder yellow by artists' colourmen almost certainly contained quercitron.[207]

9

Inorganic Reds and Purples

Red Ochre: mainly Fe_2O_3

The difference between the dull yellow, red and brown of iron oxides is difficult to define; some English writers of the past tend to group all ochres together as yellow. *Ocre de rouse* is an example of a dark ochre which, following Hilliard's description, is included amongst yellow ochres in a previous chapter even though others might regard it as red. Similarly, a number of the pigments described below as red might equally well be classed as brown.

Bole is a word which appears occasionally as a synonym for red ochre, but it is generally associated with Armenian bole, *bole Armoniack* in seventeenth-century English, a soft red ochre which was much used as a constituent in gold size and, according to de Mayerne, in canvas priming. Writers of the eighteenth century tend to mention it in passing and claim that a similar earth from the island of Lemnos was commonly sold as a substitute. Another native red ochre named after its place of origin was Indian red, a heavy pigment, more purple than red, which came from the island of Ormus in the Persian gulf. The genuine earth was highly prized, but it was hard to come by in the eighteenth century, when the pigment sold as Indian red was a manufactured iron oxide with a less purple cast. In the same way, Venetian red was once a native earth (whether found near Venice or merely imported by that route is not clear), but the name was used for a manufactured pigment in and after the eighteenth century.

Red iron oxides occur quite widely in England, Oxfordshire being one of the sources mentioned in the past. Waller suggests that red ochre is to be found at Witney; the name he uses is *reddle,* an old word for red earth which came to be used more in connection with marking sheep and cattle than with painting. The old name does appear, however, under the heading 'faint reds' in the colour list in B.M. MS. Stowe 680. The Forest of Dean in Gloucestershire was also a noted area for red ochre, and the evidence of letters patent granted in 1626 point to the fact that production was maintained on a large scale. By the terms of the grant, the patentee was given control over 'grinding and makeing that Redocker or Red Earth called Almagro and of refineing, washing, deviding from gravell or sande the burnte Ocker digged in the fforeste of Deane called Spanish Browne'.[208] The mills, vessels and drying rooms mentioned but not described in the patent confirm that earth colours, which painters themselves normally washed over in order to separate coarse particles, were in fact similarly treated on an industrial scale. Another interesting feature of the patent is the application of the Spanish word *almagro* and the name *Spanish brown* to native pigments. At one time some red and brown ochres had been imported from Spain, and, because the names were familiar to English artists, they were retained even when the pigments were produced in England. Spanish brown persisted as a colour name well into the eighteenth century, when Dossie found it necessary to point out that it was not Spanish but English.

Much red and brown ochre was prepared by calcining yellow ochre. Although the seventeenth-century patent proves that such pigments were available commercially, some painters may have prepared it themselves by the method described by de Mayerne, that is, by placing yellow ochre in a crucible on a fierce fire for at least two hours.[209] Light red came into current use as a colour name during the eighteenth century, when it was generally used to indicate a brownish red prepared by burning yellow ochre. Brown red and English red were also current at the same time, but there must be some doubt about the way in which those ochres were prepared. According to de Massoul, the yellow ochre from Berry in France was imported to Holland, where it was calcined and then sold as English red. On the other hand Watin states that English brown red is made at Deptford, seven miles from London. The proliferation of names, which were presumably introduced during the eighteenth century to cover the various nuances of colour, make it extremely difficult to determine the origin of different ochres. It is possible that some of the names mentioned above were applied not only to calcined native ochres but also to those manufactured from ferrous sulphate, that is, manufactured iron oxides.

Red ochre is frequently mentioned under one name or another in works on painting, and the remarks concerning yellow ochre are usually applied to red. The artist was advised to select the finest quality and to avoid the many coarse varieties that were more suitable for decorating than for artistic work. Red ochre was useful in oil painting grounds owing to its relatively quick drying time and its low oil absorption. In other paint layers it was often used alone, or it was mixed with lake so that a painter had a composite colour that united the brilliance of lake with the durability of ochre.

Manufactured Iron Oxide—Mars Red: Fe_2O_3

The obscure history of manufactured iron oxides in painting has already been discussed with reference to Mars yellow in an earlier chapter. Of the entire group of pigments, which includes dull varieties of yellow, orange, red, purple and brown, red was at one time the most important, because definite references to manufactured iron oxide reds appear under various traditional names in eighteenth-century literature, earlier in fact than references to yellows of similar composition. Seventeenth-century sources which are entirely devoted to painting in oils or water colours contain no recognisable references to manufactured iron oxides, although, exceptionally, Haydocke's translation of Lomazzo, published at the very end of the sixteenth century, contains a reference to burnt vitriol.

Excellent heat resistance is an outstanding feature of iron oxide red, and it is, therefore, useful in techniques involving high temperatures. Thus, references to the pigment are to be found in seventeenth-century sources which touch on the art of enamels or glass. De Mayerne's all-embracing interest in art subjects covered both topics, and one of his manuscripts includes notes on methods of employing iron to produce a red pigment for use in miniature painting in enamels. In one place de Mayerne gives detailed instructions for using steel filings to make ferrous sulphate, which may be evaporated and the residue calcined until it turns red. In another method he suggests that iron filings should be placed in an earthenware basin, covered with strong vinegar and left in the sun to dry naturally, the process being repeated about eight times in a month. The residue should be spread out quite thinly and calcined for two days, after which time it should increase in volume and turn scarlet.[210] Similar instructions are included in Merret's *The Art of Glass*.[211] De Mayerne summarised his thoughts on the subject in another passage, a translation of which is as follows:[212]

> As for red substances, they are straightforward; *crocus martis* is made quite simply with steel filings or iron chips, well washed and cleansed of all impurity, and calcined in a glass-blowers' furnace in a good earthenware dish only two inches deep, the metal spread out to a thickness of two knife handles so that the flames can penetrate it easily. Alternatively, the same saffron is made by dissolving the filings in *aqua regia*, completely evaporating the liquid, and calcining the residue for several hours until it becomes a perfect red. But the finest red of all is martial vitriol calcined to the utmost limit, ground, and washed to remove the salt, the remainder being thoroughly dried. Whatever the red, to be good for this work [enamelling] it must be tested by fire, the heat of which it must withstand without any colour change. Maybe the residue of *aqua fortis* or *caput mortuum* (as the residue left after the distillation of vitriol is called) well washed to remove the salts will leave a really excellent red substance.

It is clear that the various suggestions for making red do not correspond to later methods of making Mars colours, because there is no mention of mixing with alum and precipitating by means of an alkali. However, it is interesting to encounter a reference to *caput mortuum*, a Latin name which was used in several languages to describe the ferric oxide residue obtained as a by-product in the manufacture of fuming sulphuric acid, an industry which expanded with increasing momentum during the seventeenth and eighteenth centuries. It is clear that de Mayerne recognised the possibility of using the by-product as a pigment, and it is, therefore, quite possible that it was used in his time and during the second half of the seventeenth century, even though the earliest definite references to its use in British painting occur in eighteenth-century books.

The use of manufactured iron oxides in painting is difficult to trace because traditional names were transferred to the manufactured product. Dossie points out that Indian red was once a native earth, but, in his time, the colour sold under that name was prepared from *caput mortuum* and, in order to accommodate house-painters, it was made without the purple cast of the native earth. Scarlet ochre is the name of another manufactured variety which is mentioned by both Dossie and Field. The latter states that it is an alternative name for Venetian red, adding that it is redder and deeper than the colour sold as light red, which was calcined yellow ochre, not a wholly manufactured pigment. Two names mentioned by Field are Prussian red and English red. The first is also described by Watin as being made from *caput mortuum,* and the pigment he describes as *le brun rouge d'Angleterre* may be the same as Field's English red.

Patent literature of the late eighteenth century confirms the fact that some manufactured pigments were sold under traditional names. Atkinson's specification of 1794 (printed series number 1996) includes references to both Venetian red and Spanish brown as manufactured iron oxides. Colcothar vitriol is listed as a colour name in the same specification, where it is described as a bright, light purple. However, it was more likely a dull, brownish purple, as the same colour is referred to in the *Practical Treatise,* 1795, as 'a purple brown calx of iron'. Atkinson also lists other colours, chocolate purple brown, blue purple and dark purple brown, and proof that the patentee or a nominee actually made the colours is supplied by the inclusion of all the colour names listed by Atkinson in a sales pamphlet of the London Patent Colour Works.[213] The leaflet indicates that the pigments were used in decorators' paints, but their availability on a large scale at a relatively low price helps to explain the increasing references to manufactured iron oxides in literature on fine art in the late eighteenth century.

The tardy appearance of *Mars* as an English colour name has already been discussed in connection with yellows. Unfortunately the matter is complicated by the fact that de Massoul, who was amongst the first to use the name in England, refers to the yellow as Mars yellow but uses the Latin *crocus martis* as the name for red, explaining that it is made by calcining Mars yellow. In addition, he uses calcined vitriol as a colour name. Ackermann's tendency to follow de Massoul's colour names very closely has already been mentioned, and it is therefore not surprising to find that he uses all three names in exactly the same way in his main colour list. However, he uses the name Mars red in a subsidiary list in the treatise on water colour which he produced in 1801. There are no other references to Mars red in English literary sources of the early nineteenth century and none at all to the names Mars brown or Mars violet. Their absence suggests that the names were not in current use in English before 1835.

According to manufacturers, such as de Massoul, an outstanding feature of any man-made iron oxide was its fineness, that is, it had smaller, more homogeneous pigment particles than were to be found in native ochres. De Massoul states uncompromisingly that reds made from martial vitriol are superior to calcined ochres. The difference between native and manufactured varieties may account for the fact that, whereas eighteenth-century writers on painting warn that the coarse types of native ochre used by decorators should be avoided, painters were at the same period prepared to employ the manufactured varieties which were supplied to decorators and artists alike, first under the guise of traditional colour names and later under special names which indicated that the pigments were modern, manufactured products.

Red Lead: Pb_3O_4

The orange-red pigment produced by prolonged heating of lead was generally known as red lead, although *minium* (or the variation *minne*) was a synonym in current use during the seventeenth century. One or two writers of that period who had studied classical authors were confused by Latin nomenclature, and, consequently, Peacham states that *minium* means vermilion and Waller declares that red lead was unknown to the ancients. It is now well known that red lead and the associated pigments, white lead and yellow lead monoxide, have a very long history and that, although Pliny used the name *minium* for cinnabar (native vermilion) and *minium secundarium* for red lead, *minium* was a generally accepted name for red lead during the sixteenth century and later.[214]

The pigment could be made in small quantities by roasting white lead over a fire, but it held considerable importance as a decorators' paint and it was therefore manufactured on an industrial scale. The following description of the process was printed in England in the seventeenth century.[215]

> First they take Lead and waste it in an Oven or Furnace: that is bring it to a Substance almost like a Litharge, by stirring it with an iron rake or how. This they grind with two pair of stones which deliver it from one to another: the first grind it courser, the second finer. (There is a mill so contrived as that it moves at once six pair of these stones.) Thus reduced to powder and washed it is put into an Oven or reverberating Furnace, and, by continual stirring with the iron rake or how, it is brought to the right colour in two or three dayes. The fire must not be extreme all this while, else it will clod together and change colour. The iron rake wherewith it is stirred is hung or poised on an iron hook, else it is so heavy that it could not be moved by one man.

During the eighteenth century the pigment was made from litharge at Holywell in Flintshire and by direct calcination of lead in Derbyshire,

Fig. 38. Red lead manufacture during the first half of the nineteenth century, showing the suspended rod which enabled the workman to stir the lead as it was being roasted in the furnace (Dodd, *British Manufactures: Chemical*, 1844). (Reproduced by courtesy of the University of London Library.)

where there were nine red lead mills. These details are mentioned by Watson, who also states that manufacture of the pigment was well understood in Britain and Holland but not in France.[216] Some writers of the late eighteenth century suggest that red lead was imported from Holland, but, even so, manufacture continued in England during the nineteenth century, for there is an illustration from that period (*Figure 38*) showing the operation of the suspended rake or stirring rod described almost two centuries earlier.[217]

Red lead is mentioned quite frequently in literature on painting, although it is not recommended as highly as some other colours. Bate comments on its quick-drying properties in oil, but de Mayerne states that it lacks permanence and, moreover, it should not be used in priming. It has nevertheless been identified, sometimes in association with red ochre, in oil painting grounds of this period.[218] Norgate includes it in his list of colours for miniature painting but takes it as his example for describing the thoroughness with which certain pigments must be washed over before use. Towards the end of the eighteenth century some writers thought that red lead was going out of use, but Williams, writing of oil colours, recommends an improved red lead which is bright and permanent, and de Massoul lists saturnine red, describing it as an expensive variety of *minium* which is purified by repeated washing with distilled water. The conviction that the durability of red lead had been improved in some way must have been widespread, because, apart from its cheapness which might have made it attractive to some painters, there is otherwise no reasonable explanation for its inclusion in a large number of books on water-colour painting written during the early nineteenth century. Samples of the colour are to be found in the books by

Roberts and by Pretty and in the anonymous work *Le maître de miniature*. However, the French writer mentions that red lead is liable to discolour quickly, so it is possible that his sample was included as a warning, for the area which should have been bright orange originally is now partially grey.

Realgar: As_2S_2

This orange-red pigment is closely associated with orpiment, another sulphide of arsenic, and is usually found with it, although in smaller quantities, in the same deposits. On the rare occasions when it is mentioned in early English books on painting, it is referred to as red orpiment. It is listed under that name in V. & A. MS. 86.EE.69 and also in B.M. MS. Sloane 6284, where it is listed with browns. Haydocke calls it burnt orpiment and describes it as 'orange tawnie'. The fact that realgar was at one time known as orpiment helps to explain why no references to it were found in the customs ledgers, for it is quite likely that yellow and red sulphide of arsenic were imported together under one name. Realgar has similar chemical properties to the yellow pigment in that it is incompatible with others containing lead or copper, and it is probably for that reason that it was not used much in painting.

Dossie and some other writers of the eighteenth century make no mention of realgar at all. Field, who calls the pigment orange orpiment and gives realgar as a secondary name, states that there are native and manufactured varieties. No one else refers to the pigment as manufactured, but it is possible that there was a slight revival of the use of realgar in the early nineteenth century, as a colour sample is included in the book by Roberts.

Vermilion: HgS

Red mercuric sulphide occurs naturally and it has been manufactured in Europe for use as a pigment since the early medieval period.[219] In English two names, cinnabar and vermilion, have been used interchangeably in the past to describe either the natural or the manufactured product, but, by the seventeenth century, vermilion was used more frequently. Browne speaks of artificial vermilion and native cinnabar, but it does not seem to have been until the second half of the eighteenth century that cinnabar was applied only to the unground, native mineral.

Thompson, writing of a much earlier period, stresses the very great value of the cinnabar which was mined in Spain during the Roman period when people in the western world did not known how to make vermilion, and he also explains its important role in medieval painting when it was the best bright red available.[220] However, it was more moderate in price by the seventeenth century; the price list in MS. Stowe 680 includes it at sixteen pence per pound, admittedly not the cheapest

Fig. 39. Manufacture of vermilion in the sixteenth century. Pots containing black mercuric sulphide were arranged in an open hearth, logs were placed over them and set alight. The contents of the pots sublimed and condensed in the form of red mercuric sulphide (Agricola, *De re metallica*, 1556). (Reproduced by courtesy of the University of London Library.)

colour but not amongst the most expensive either. The manuscript sources which contain medieval recipes for pigments or drugs generally include instructions for making vermilion (the fifteenth-century B.M. MS. Sloane 122 contains more than one), but later documentary sources include the information much less frequently and so point to the fact that the pigment was then easily obtainable from colour shops. Towards the end of the seventeenth century, however, the complaint that the pigment was adulterated with red lead prompted some writers to suggest that a painter should either acquire the native mineral in lump form or, alternatively, that he should make it himself. John Smith gives the following instructions:[221]

> Take six Ounces of Brimstone and melt it in an Iron-Ladle, then put two Pound of Quicksilver into a Shammy Leather, or double Linnen-Cloth, squeeze it from thence into the melted Brimstone, stirring them in the mean time with a wooden Spatula, till they are well united, and when cold, beat the mass into a Powder, and sublime it in a glass Vessel, with a strong Fire, and it will arise into that red substance which we call artificial Cinaber, or Vermillion.

The proportions of sulphur and mercury vary enormously from one recipe to another. In Chambers' *Cyclopaedia* 1728, three parts of sulphur to four of mercury are suggested, whereas one part of sulphur to seven of mercury are recommended in the instructions for the Dutch method given in *Annales de chimie,* where it is described as the method by which the Dutch produced the best vermilion in Europe. A proportion of about one part sulphur to three of mercury is mentioned in the account of Pekstok, a seventeenth-century Dutch manufacturer.[222]

Despite Smith's recommendation, it is doubtful if many painters of the seventeenth and eighteenth centuries made their own vermilion, unless they were also amateur chemists, because the process of sublimation, which was essential to turn the substance from black to red, required equipment which only a chemist was likely to have. Dossie explains in *The Handmaid to the Arts* that sublimation is the raising of a solid body in fumes which are then condensed, and he specifies that special, spherical glass vessels called cucurbits must be obtained from the glass-blowers. The process of manufacture was tedious even with the right equipment, as the vessels required special preparation so that the lower half could be suspended in a furnace and they needed constant attention to ensure that the vermilion which condensed inside the upper part of each sublimer did not close the mouth of the vessel. Finally, the only way of removing the pigment was to break the glass.

In view of the trouble involved in making vermilion, Dossie's instructions for detecting the presence of red lead adulteration were probably more useful than details of manufacture. He suggests that a very orange colour points to the presence of red lead, explaining that it can be detected by heating the pigment together with charcoal dust in a crucible on an ordinary fire. After some time, any lead which the vermilion contains will be found at the bottom, and the proportion of lead adulterant can be determined by comparing the weight before and afterwards. Unfortunately, the adulteration of vermilion was a widespread practice in England, and English vermilion had a poor reputation abroad for that reason. Watin gives a specific warning that true vermilion and English vermilion are not the same. According to most accounts, the Dutch product was good and Chinese vermilion was also of high quality. The eighteenth-century customs records show that Dutch vermilion was imported regularly and in increasing quantities until it reached a peak of just under 32 000 pounds in 1760. After that time it decreased as greater quantities were imported from China and Germany. Chinese vermilion was imported occasionally during the first half of the eighteenth century but its reputation was not firmly established until the second half of the century when many writers recommend it. By that time it was profitable to pass off other vermilion as Chinese; Payne relates how unscrupulous salesmen would make up packets of fourteen ounces and mark Chinese characters on the outside in order to mislead the purchaser.

The frequent adulteration of vermilion must be taken into account when considering comments on the pigment, as the addition of red lead or an organic red could affect its permanence. Pure vermilion may occasionally revert to black, a failing which de Massoul mentions, but earlier writers do not explain if that is what they mean when they say that it is liable to change. De Mayerne states that vermilion is useless in oil as it changes and affects other colours, and Elsum also criticises it on the grounds of alterability and poor drying. On the other hand, numerous other writers, including Bate, Gyles, Smith, Bardwell and Williams, appear to have been satisfied with it and list it amongst the colours generally used in oil painting. As one might expect, early writers who were concerned with limning in the medieval tradition cite vermilion as the finest red of all. The author of *Limming*, 1573, and Peacham recommend that it should be tempered with glair. However, as vermilion is opaque and it generally has relatively large pigment particles, it was not particularly useful in portrait miniatures; consequently Norgate and other seventeenth-century writers deliberately excluded it from their colour lists. Surprisingly, its opacity did not deter the painters who coloured prints and maps, and writers of the second half of the seventeenth century often recommended it for that purpose although a warning that it must be well ground is usually included. During the following century it came back into favour amongst miniature painters who, according to many writers, used it for drapery, a comment which suggests that the popularity of red military uniforms had something to do with its reinstatement. In a letter dated 1814, Chinnery remarked in connection with miniature painting: 'There is something about Vermilion very curious—Vermilion is Vermilion—but Newman does contrive to make *his* cakes of a very different Color to all others . . . it is the only colour which does for Coats at all.'[223] The pigment was favoured by flower painters during the nineteenth century and it was used for local colour in many other types of painting. As Varley points out, it is an extremely brilliant colour which has the effect of making other colours, even other reds, recede in a painting. A wash of vermilion is included in the work of Varley and Fielding, and a sample of orange vermilion, made according to Field's own undivulged method, is included in *Chromatography*.

Gold Purple

The purple-red pigment known as gold purple or purple precipitate of Cassius came to the attention of scientists in the late seventeenth century. Andreas Cassius the elder, a physician working in Hamburg at that time, was credited with its discovery, although a modern authority states that red glass of similar composition was described in the fourteenth century.[224] The earliest English reference to gold purple as a pigment is Dossie's description of its manufacture for use in enamelling. According to that writer, the pigment, which was more crimson than purple, was made by dissolving powdered gold and tin filings separately in *aqua regia* (a mixture of nitric and hydrochloric acid). First, some of the gold solution was poured into a quantity of water and then a smaller proportion of tin solution was added so that a red precipitate began to form; small quantities of both solutions were added alternately until both were used up. Finally, the clear fluid was separated from the red precipitate, and the pigment was then washed with spring water and dried. To produce a colour which was much nearer purple, the same method was followed with a solution of gold to which salt of tartar (potassium carbonate) had been added. Dossie thought of gold purple as an enamel only, but, by the end of the century, precipitate of gold was used as an artists' colour, as it appears under that name in de Massoul's list. An example is included in Field's 'Practical Journal 1809', f. 361 (see *Plate 4*). According to Field's comments in *Chromatography,* the colour was very durable but was restricted to use in miniature painting as it was so costly. Purple madder was recommended as an acceptable and much cheaper alternative.

Iodine Scarlet: HgI_2

A scarlet pigment, iodide of mercury, the hue of which was in some ways comparable with vermilion, was manufactured a few years after the discovery of iodine by a French manufacturer of saltpetre in 1811 or 1812. Sir Humphry Davy acquired some iodine on a continental trip and, in 1814, after his return to England published the results of his research, proving that it was an element.[225] In the same year, Vauquelin, who is frequently credited with the discovery of the element, published information concerning various compounds, including iodide of

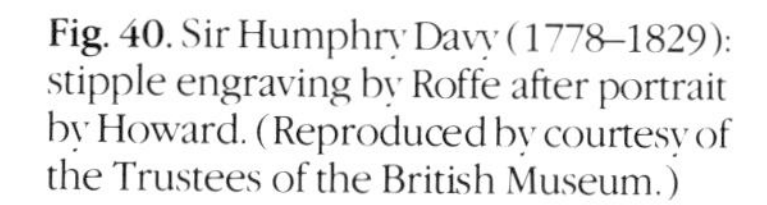

Fig. 40. Sir Humphry Davy (1778–1829): stipple engraving by Roffe after portrait by Howard. (Reproduced by courtesy of the Trustees of the British Museum.)

mercury. Its preparation was very simple; if iodine and mercury were ground together in a stone mortar, they quickly united and formed a red substance.[226] More detailed instructions for preparing a solution of iodide of zinc and mixing it with chloride of mercury are included in Mérimée's book, which also contains the statement that in England the pigment is sold as scarlet lake.

Four samples of the colour are included towards the end of Field's 'Practical Journal 1809'. Two of them are named simply scarlet, one of these having been obtained 'of Mr Hobday from Mr Sheffield's Painting box' and the other from the artist James Ward. The latter sample Field levigated to produce another specimen listed as orange, and the fourth sample is listed as iodide of mercury. In his notebook Field remarks upon the rapidity with which the colour faded when tested in sunlight and in damp, impure air, and comments '... gradually and totally faded in the shade, as if eaten away at the edges. In this book these colours are gradually vanishing'.[227] It is a fact that, despite having been protected from light for many years, the more thinly applied areas of iodine scarlet in Field's notebook have faded completely and the only vestiges of colour remaining are in areas where the water colour was applied in greater concentration (see *Plate 5*). The same effect is to be seen in an example in a Winsor & Newton tint book dating from the mid-nineteenth century.

In *Chromatography,* Field calls the pigment iodine scarlet, describes it as having the body and opacity of vermilion, and warns that it is not permanent. The name *iodine scarlet* does not appear in any of the colourmen's catalogues examined; it is evident that the artists' colour was sold under the name *pure scarlet*, while *iodide mercury* is the name used in Berger's manuscript documents.[228] The colour was more expensive than either scarlet vermilion or scarlet lake, which, according to Field, was prepared from cochineal and often adjusted with vermilion.

Chrome Red: $PbCrO_4 \cdot Pb(OH)_2$

Another new red, which, like iodine scarlet, did not become a permanent addition to the artists' palette, was basic lead chromate. When Vauquelin first discussed chrome compounds in 1809, he pointed out that various nuances of colour could be obtained. If potassium chromate with an excess of alkali was used to make lead chromate, a yellow-red or sometimes a fine, deep red would be obtained.[229] From time to time chrome red was mentioned in scientific papers. One which appeared in *Annals of Philosophy* in 1825 refers to methods of manufacture published in French in 1812 and 1822. The earlier one consisted of boiling potassium chromate and basic lead carbonate together. The later method was to boil chrome yellow (lead chromate) with potash.[230] It was then regarded as a permanent pigment in oil and was recommended as being superior to red lead. Chrome scarlet is listed in one of Winsor & Newton's early catalogues, *circa* 1840–42, but it is not to be found in later editions.

Colour Plates

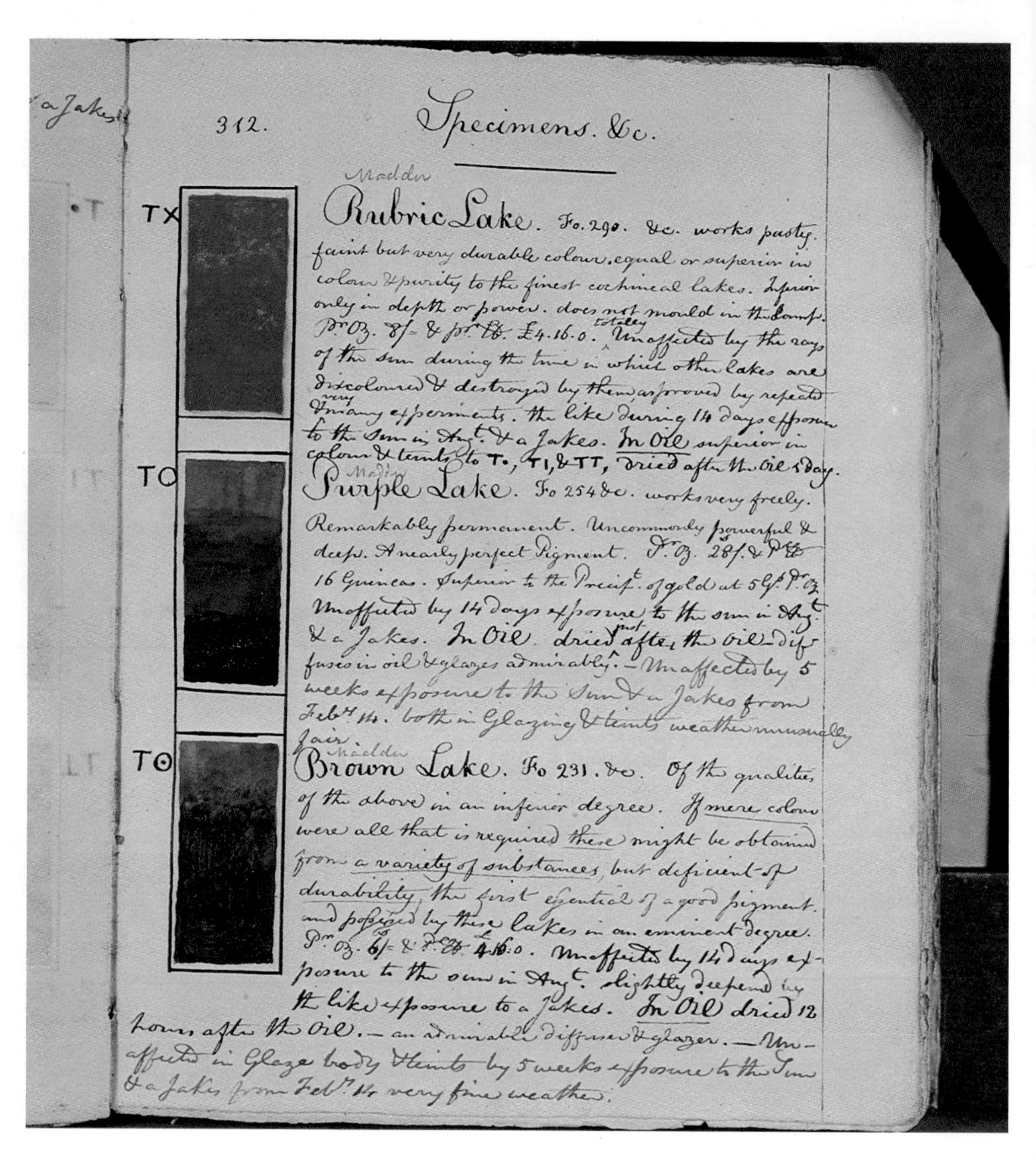

312. Specimens. &c.

TX Madder Rubric Lake. Fo. 290. &c. works pasty. faint but very durable colour. equal or superior in colour & purity to the finest cochineal lakes. Inferior only in depth or power. does not mould in the Comp. Pr Oz. 8/- & pr lb. £4.16.0. totally Unaffected by the rays of the sun during the time in which other lakes are discoloured & destroyed by them, as proved by repeated & very many experiments. the like during 14 days exposure to the Sun in Augt. & a Jakes. In Oil superior in colour & tints to T., T1, & TT, dried after the Oil 5 Day.

TO Madder Purple Lake. Fo 254 &c. works very freely. Remarkably permanent. Uncommonly powerful & deep. A nearly perfect Pigment. Pr Oz. 28/. & Pr lb 16 Guineas. Superior to the Precipt. of gold at 5 G. Pr Oz. Unaffected by 14 days exposure to the sun in Augt. & a Jakes. In Oil dried not after the Oil — diffuses in oil & glazes admirably. — Unaffected by 5 weeks exposure to the Sun & a Jakes from Feby 14. both in Glazing & tints weather unusually fair.

TO Madder Brown Lake. Fo 231. &c. Of the qualities of the above in an inferior degree. If mere colour were all that is required these might be obtained from a variety of substances, but deficient of durability, the first essential of a good pigment and possessed by these lakes in an eminent degree. Pr Oz. 6/- & Pr lb £4.16.0. Unaffected by 14 days exposure to the sun in Augt. slightly deepened by the like exposure to a Jakes. In Oil dried 12 hours after the Oil. — an admirable diffuser & glazer. — Unaffected in Glaze body & tints by 5 weeks exposure to the Sun & a Jakes from Feby 14 very fine weather.

Plate 1 Three examples of madder lakes made by George Field. The symbols beside the samples are a numerical notation and the text contains references to earlier notebooks. Field, 'Practical Journal 1809', Courtauld Institute of Art MS Field/6, f.312

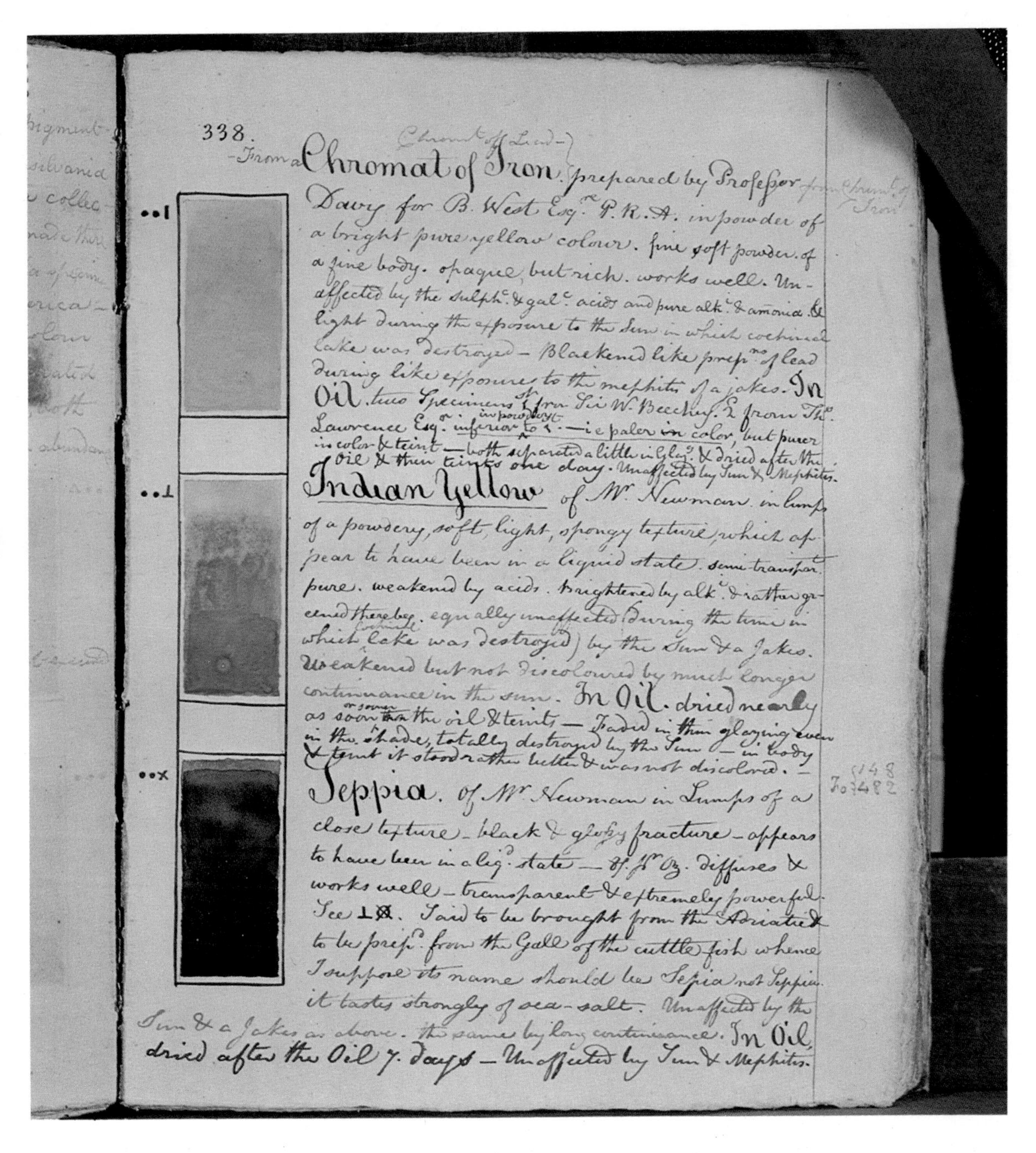

338.

(Chromat of Lead)

— From a Chromat of Iron. prepared by Professor Davy for B. West Esq.re P.R.A. in powder of a bright pure yellow colour. fine soft powder of a fine body. opaque, but rich. works well. Unaffected by the sulph.c & gal.c acids and pure alk.s & amonia & light during the exposure to the Sun in which cochineal lake was destroyed — Blackened like prep.ns of lead during like exposures to the mephitis of a jakes. In Oil. two Specimens 1st from Sir W. Beechey. 2 from Th.s Lawrence Esq.r in powder inferior to 1. — i.e. paler in color, but purer in color & teint — both separated a little in glaz.g & dried after the Oil & then teints one day. Unaffected by Sun & Mephitis.

Indian Yellow of Mr. Newman in lumps of a powdery, soft, light, spongy texture, which appear to have been in a liquid state. semi-transparent pure. weakened by acids. brightened by alk.s & rather greened thereby. equally unaffected (during the time in which cochineal lake was destroyed) by the Sun & a Jakes. Weakened but not discoloured by much longer continuance in the sun. In Oil. dried nearly as soon or sooner than the oil & teints — Faded in thin glazing even in the shade, totally destroyed by the Sun — in body & teint it stood rather better & was not discolored. —

Seppia. of Mr. Newman in Lumps of a close texture — black & glossy fracture — appears to have been in a liq. state — Sp. Gr. Oz. diffuses & works well — transparent & extremely powerful. See 1X. Said to be brought from the Adriatic & to be prep.d from the Gall of the cuttle fish whence I suppose its name should be Sepia not Seppia. it tastes strongly of sea-salt. Unaffected by the Sun & a Jakes as above. the same by long continuance. In Oil dried after the Oil 7. days — Unaffected by Sun & Mephitis.

Plate 2 Chrome yellow, Indian yellow, sepia. The name 'Chromat of Iron' was later corrected in pencil to 'Chromat of Lead'. It is an early example of a relatively new pigment. Benjamin West supplied the sample; Beechey and Lawrence also had the pigment. Samples of Indian yellow and sepia came from the London colourman, Newman. Field, 'Practical Journal 1809', Courtauld Institute of Art MS Field/6, f.338

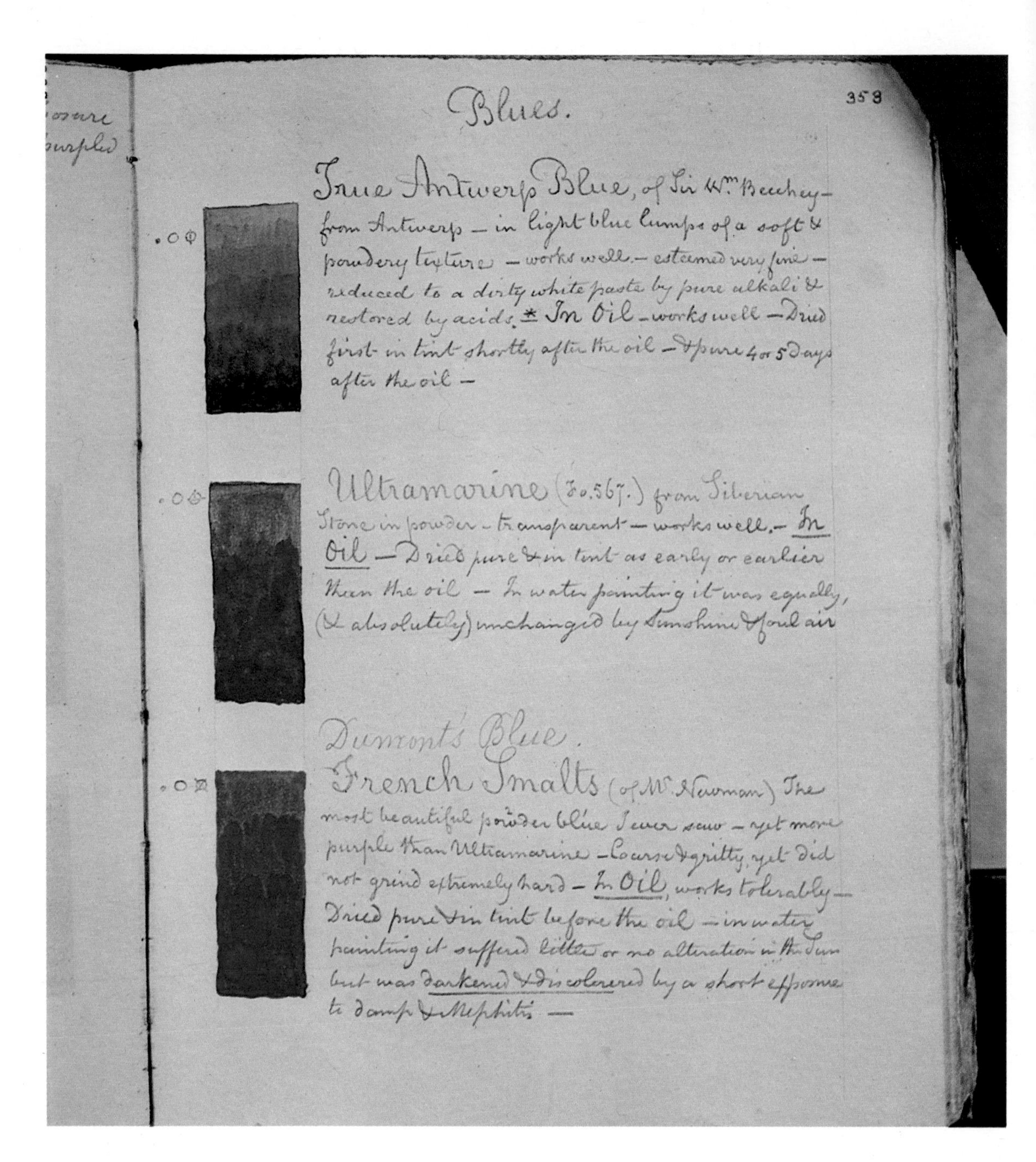
353

Blues.

True Antwerp Blue, of Sir Wm Beechey — from Antwerp — in light blue lumps of a soft & powdery texture — works well. — esteemed very fine — reduced to a dirty white paste by pure alkali & restored by acids. * In Oil — works well — Dried first in tint shortly after the oil — & pure 4 or 5 Days after the oil —

Ultramarine (No. 567.) from Siberian Stone in powder — transparent — works well. — In Oil — Dried pure & in tint as early or earlier than the oil — In water painting it was equally, (& absolutely) unchanged by Sunshine & foul air

Dumont's Blue.

French Smalts (of Mr Newman) The most beautiful powder blue I ever saw — yet more purple than Ultramarine — Coarse & gritty, yet did not grind extremely hard — In Oil, works tolerably — Dried pure & in tint before the oil — in water painting it suffered little or no alteration in the Sun but was darkened & discoloured by a short exposure to damp & Mephitis —

Plate 3 Antwerp blue, natural ultramarine, smalt. This page from Field's 'Practical Journal 1809' gives an idea of the variety of blue pigments available then. Courtauld Institute of Art MS Field/6, f.353

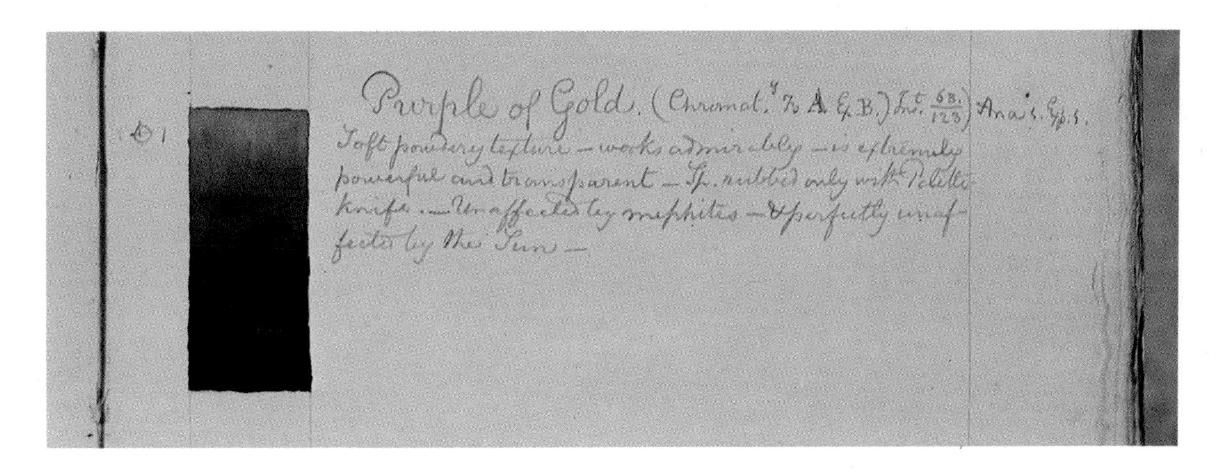

Purple of Gold. (Chromat.y To A & B.) Int. 68/123) Ana's. Ep.s.
Soft powdery texture — works admirably — is extremely powerful and transparent — Sp. rubbed only with Pallette knife. — Unaffected by mephites — & perfectly unaffected by the Sun —

Plate 4 Gold purple, sometimes called precipitate of gold or purple of Cassius, was a rare pigment because of its high cost. Field, 'Practical Journal 1809', Courtauld Institute of Art MS Field/6, f.361

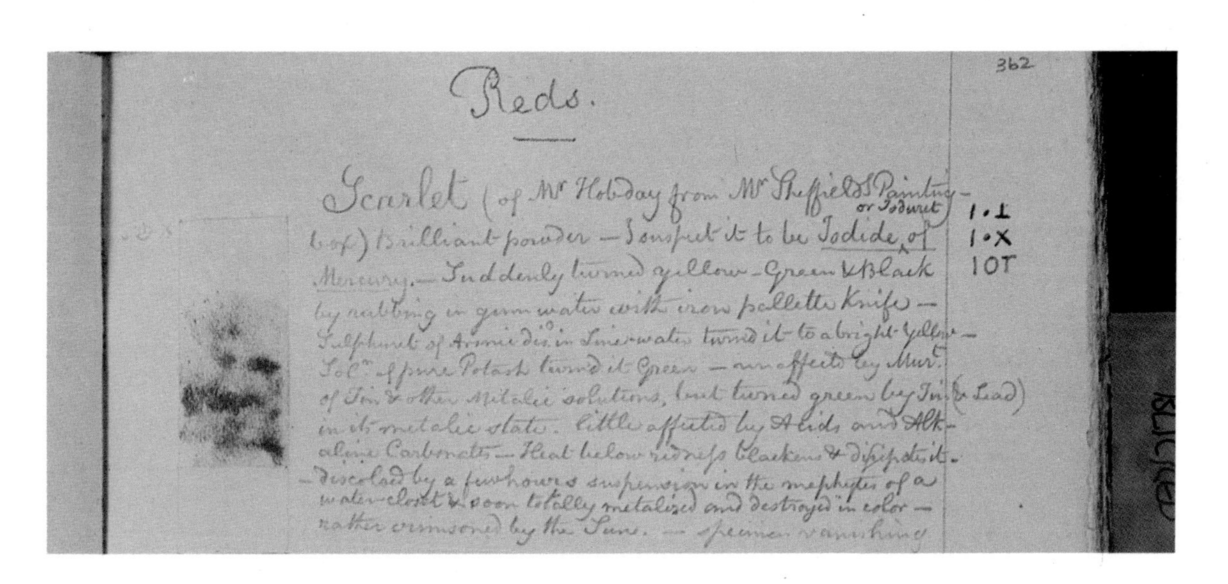

362

Reds.

Scarlet (of Mr Hobday from Mr Sheffield's Painting — or Ioduret of) Brilliant powder — I suspect it to be Iodide of Mercury. — Suddenly turned yellow — Green & Black by rubbing in gum water with iron pallette knife — Sulphuret of Ammonia in Lime-water turned it to a bright yellow — Sol.n of pure Potash turned it Green — unaffected by Mur.t of Tin & other Metalic solutions, but turned green by Tin (& Lead) in its metalic state. Little affected by Acids and Alkaline Carbonates — Heat below redness blackens & dissipates it. — discolored by a few hours suspension in the mephytes of a water-closet & soon totally metalised and destroyed in color — rather crimsoned by the Sun. — specimen vanishing

1.1
1.X
1OT

Plate 5 Iodine scarlet, sometimes sold as scarlet lake or pure scarlet. This example demonstrates the fugitive nature of the pigment that was introduced and became obsolete in the nineteenth century. Field, 'Practical Journal 1809', Courtauld Institute of Art MS Field/6, f.362

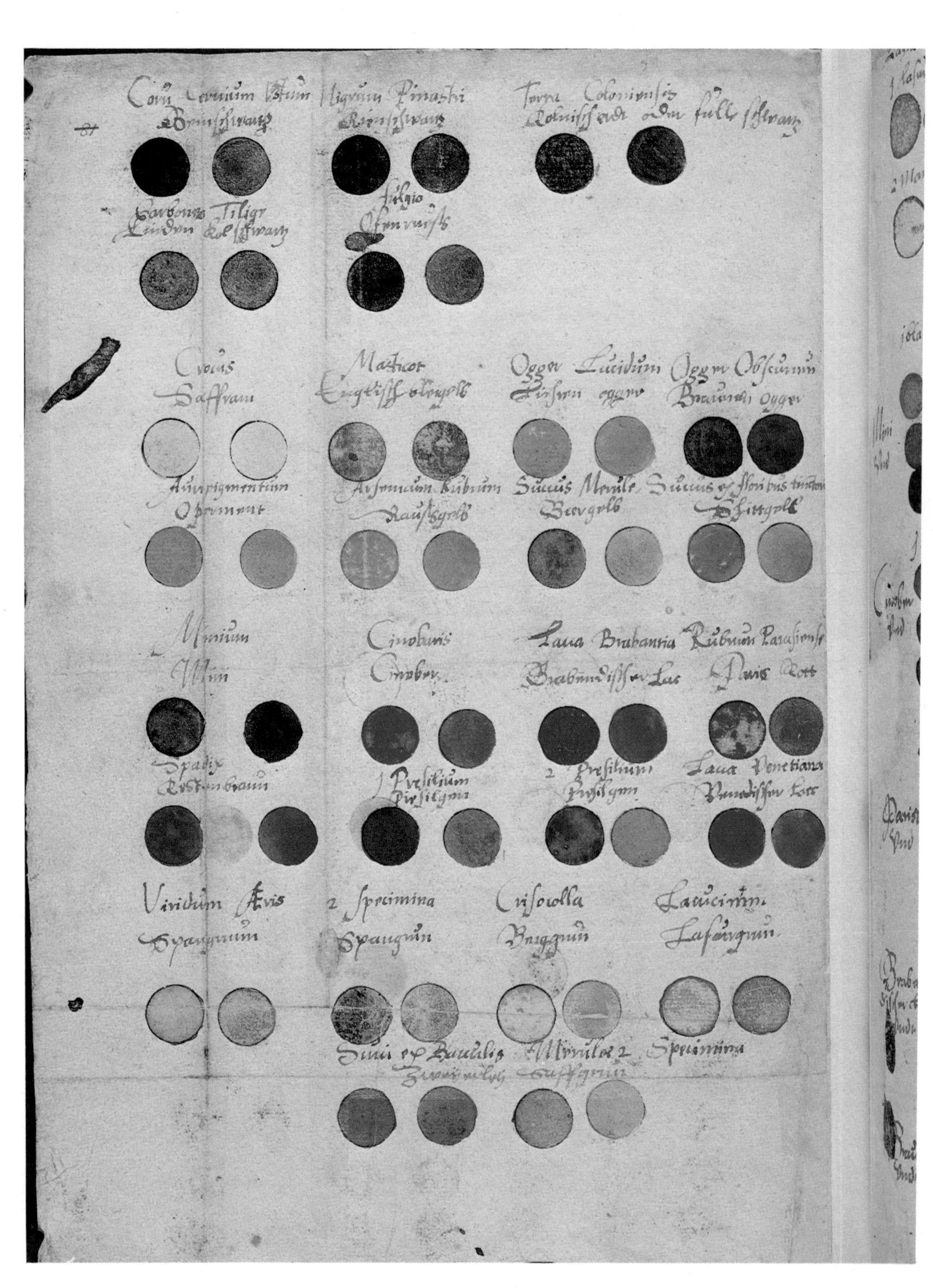

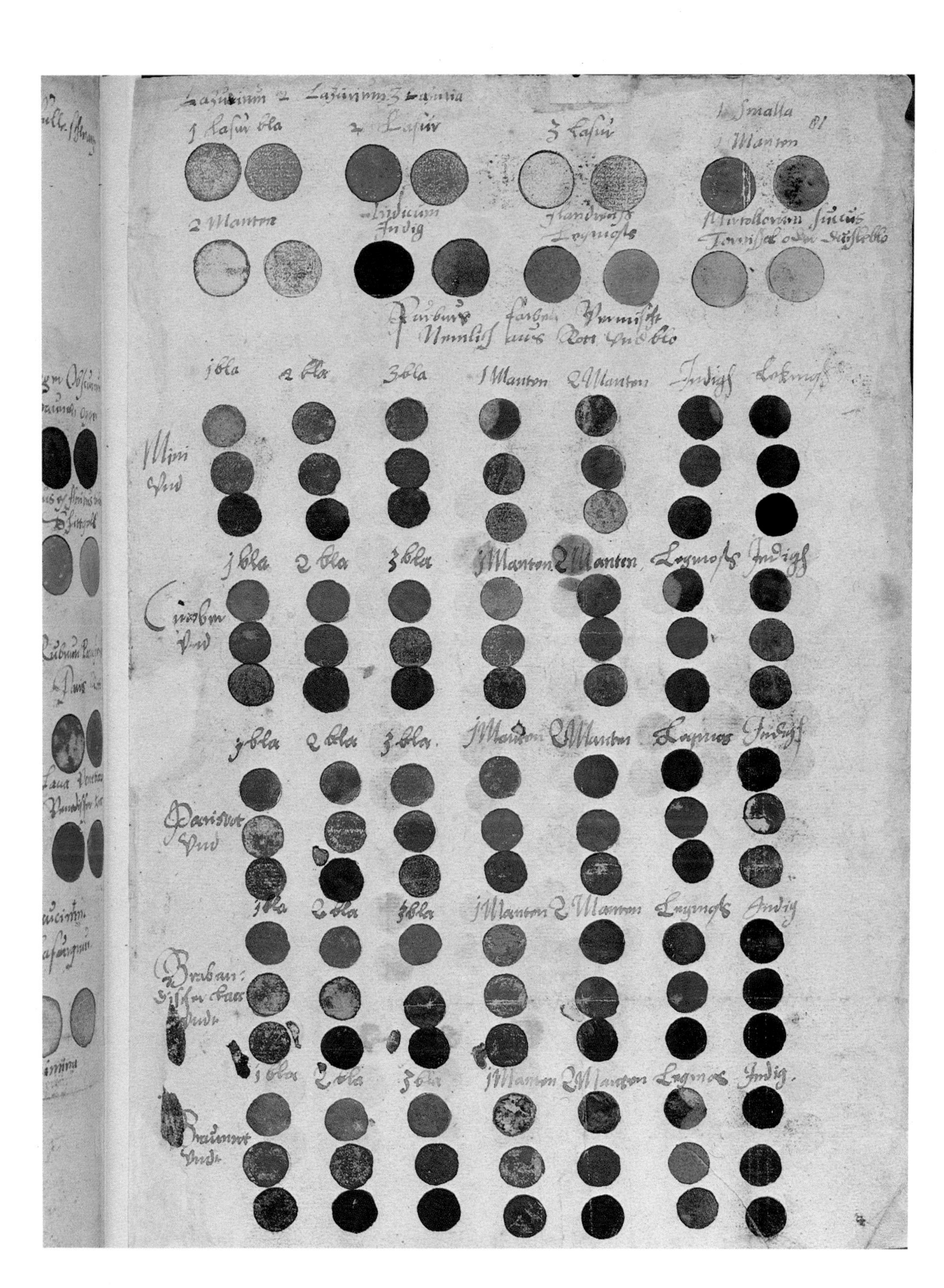
81
Smalta
1 bla
2 bla
3 bla
1 Manton
2 Manton
Indig

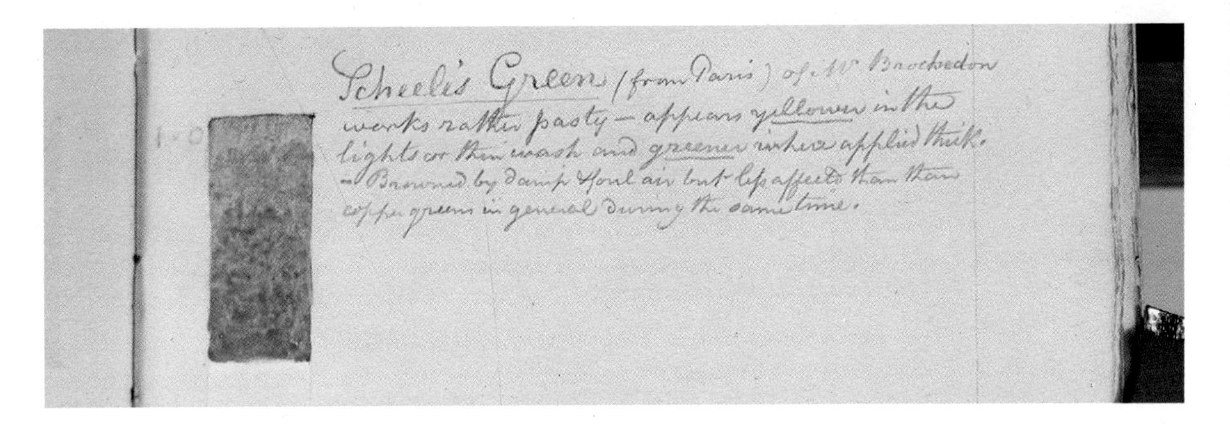

Scheele's Green (from Paris) of Mr Brockedon
works rather pasty — appears yellower in the
lights or thin wash and greener where applied thick.
— Browned by damp & foul air but less affected than other
copper greens in general during the same time.

Plate 6 Scheele's green from Field's 'Practical Journal 1809', Courtauld Institute of Art MS Field/6, f.372

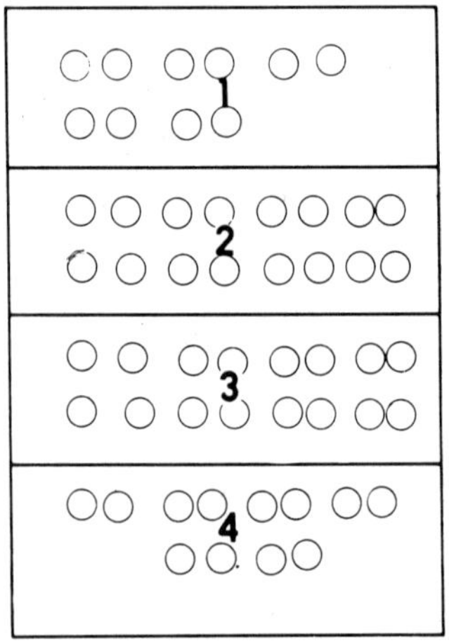

Plate 7 (preceding pages, left)

Samples of water colours dating from the first half of the seventeenth century. One of four pages of samples in the Mayerne manuscript, British Museum MS Sloane 2052, f.80v. For comments on the colours please refer to Appendix 2, page 193

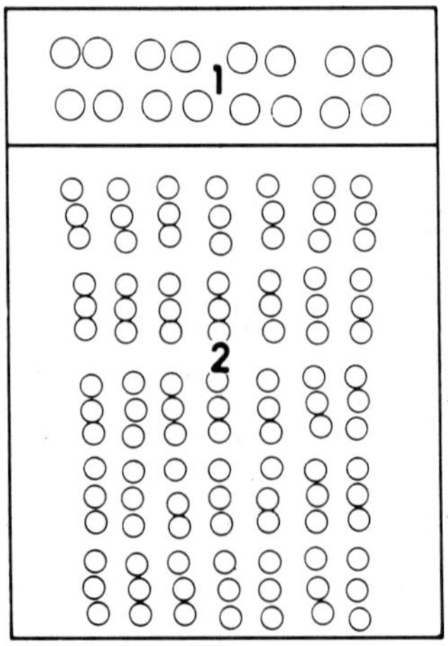

Plate 8 (preceding pages, right)

Samples of blue water colours and purple mixtures from the Mayerne manuscript, British Museum MS Sloane 2052, f.81. For comments on the colours please refer to Appendix 2, page 194

10

Organic Reds

The complicated history attached to both the nomenclature of lake colours and their manufacture has already been mentioned in connection with the yellow pigment called pink. In the early seventeenth century the name *lake* used on its own without qualification always indicated a red pigment made by what is now known as the lake process, that is, extracting a dye and precipitating it on a base such as aluminium hydroxide. At the beginning of the century *lake* was not restricted to the colour made from lac, the substance with which it was originally associated, but it could be used for a red pigment prepared not only from lac, but also from cochineal, kermes, madder, brasil wood, or a combination of several of these materials.[231] As a result, unless the pigment is described or the name is qualified in documentary sources, it is often impossible to identify the raw material from which the colour was obtained. Owing to the often vague references to red lakes and their manufacture in early sources, it is difficult to discuss the history of the different pigments in the same detail. Consequently, the animal substances are discussed under the headings Indian Lake and Carmine; references to past confusion over names for reds are made in connection with Indian lake, while comments on manufacture are included under carmine. The vegetable substances are discussed under the headings Brasil and Madder.

Indian Lake

Lac is obtained from the females and eggs of the insects known as *Coccus lacca,* which infest various trees, especially fig trees, indigenous to Asia and India. They, and kermes and cochineal insects, are of a type commonly called 'shield-louse', as they are small and round with a shield over the back. Female lac insects have vestigial wings and legs and spend their whole life gathered in large clusters on host plants. When Indian lac is harvested, its animal origin is virtually unrecognisable, because it is a solid substance made up of the bodies of female insects which are dead, each with some 200 to

500 unhatched eggs, all surrounded by a brown-red, hardened exudation. Individual insects are not visible, and the whole substance looks like some form of growth on the host plant. It is collected by breaking off lac-bearing branches before the larvae hatch. Some lac is left, and at swarming time the larvae of minute size can be seen for a few days moving about to find a place on the tree to settle and feed.[232] In ancient times Asiatics, who were able to observe the life cycle, knew that lac was a substance of animal origin, so various names meaning 'little worm' were developed for lacca and similar insects. To Europeans, however, lac appeared to be part of the sticks on which it was imported, thus the name *coccus* (berry or acorn) was attached to lac and later to the live insects, a misnomer for which the ancient Greeks and Romans are held responsible.[233]

Discussion of the confused nomenclature connected with a number of red pigments has been undertaken by other writers. The origin of the links between Latin *coccus* and *granum* and English grain (all of which were associated with the idea that lac was a seed or berry), and the links between *kermes* and *vermiculum* (both meaning little worm) are to be found during the medieval period and they are ably discussed by Thompson.[234] When continuing the discussion with special reference to seventeenth-century England, it is possible to ignore the name grain, which, apart from its specialised use in textile dyeing, was obsolete. Nevertheless, the seventeenth century presents an additional problem in that the words *cinnabar* and *sinopia* were wrongly associated with the name *sinoper lake,* which may have been a development of the medieval name cynople (Latin *sinopia*) mentioned by Thompson.

Sinoper lake and the variation *topias* are to be found in sources dating from the late sixteenth and early seventeenth centuries.[235] The composition of both is uncertain; Thompson states that the medieval cynople was a composite lake, and it seems that sinoper may have been similar, for that mentioned in B.M. MS. Sloane 1394 was derived from the dye in scarlet cloth. (For quotation see the section on carmine.) On the other hand, the price lists in V. & A. MS. 86.EE.69 and B.M. MS. Stowe 680 show that sinoper lake was very much more expensive than sinoper topias, so much so that it is reasonable to suppose that the former might have been made from lac or cochineal, both of which were extremely costly at that time. The word topias is an obsolete variation of topaz, a circumstance which suggests that the cheaper colour tended towards orange. The name is, however, always listed with other reds, and in MS. Stowe 680 it is included with crimson colours. By the second half of the seventeenth century the name sinoper was no longer current, and Browne, who was one of the few to refer to it, misguidedly tried to modernise the spelling, listing it as cynnabar lake. Confusion over cinnabar and sinoper is understandable, as the names sound very familiar. Erratic spelling was a common failing amongst writers of the sixteenth and seventeenth centuries, and it is, therefore, not surprising to encounter a recipe for vermilion (or cinnabar) headed 'For to make Synoper' in Bodleian MS. Ashmole 1480, f. 12. Doubtless most writers of the early seventeenth century knew what they meant, but Peacham was hopelessly muddled, for he confused sinoper not only with cinnabar but also with sinopia, saying all of them were made from brimstone and quicksilver. *Sinopia* is not the equivalent of vermilion but is the classical Latin name used for a special, red iron oxide. The danger of attempting to interpret the nomenclature of one period by reference to another far removed from it, in this case by reference to classical antiquity, is all too apparent from the inaccurate statement concerning sinoper which appears in the first edition of Peacham's book *The Art of Drawing with the Pen.* However, in the edition of 1612 the statement was revised as follows:[236]

> Sinaper . . . hath the name Lake of *Lacca,* a red berry, whereof it is made growing in China and those places in the East Indies, as Master Gerrard shewed me out of his herbal, it maketh a deepe and beautiful red, or rather purple, almost like vnto a red Rose.

Peacham's reference to Gerard's description of Indian lac as a berry may have helped to perpetuate a misconception which was fairly common in seventeenth-century England. Even so, it would be wrong to give too much importance to the statements of those who thought of lac as a berry, because there were at the same time others who were well aware of the true nature of lac. For example, Parkinson's herbal contains an explanation that lac is formed by 'winged ants' which settle on trees, that the substance is sometimes imported on sticks (stick lac), or, after it has been cleared from the sticks and melted, it can be imported in cake form, or in thin pieces (shellac).[237] In order to amplify Parkinson's description, it is worth adding a few comments

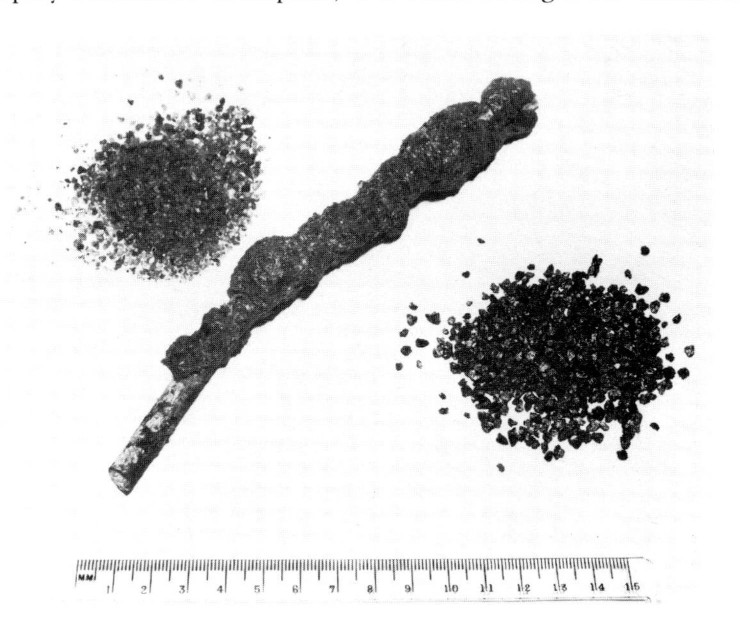

Fig. 41. Indian lac and cochineal insects. In the centre is a piece of stick lac and, above it, some seed or grain lac that has been stripped from a branch. On the right is a group of dead cochineal insects. (Reproduced by the kind permission of Messrs Winsor and Newton Limited.)

taken from a modern authority. When the dead insects are stripped from the branch and crushed, the seed or grain lac, as it is then called, is immersed in hot water to separate the grains from the colouring matter, the liquid is evaporated, and the residue which remains may then be formed into cakes for use as a dye. It was worth importing untreated stick lac, because it was capable of providing a greater quantity of colouring matter than the cakes.[238] Eighteenth-century customs records support this view, for, in 1700, 25 543 pounds of stick lac and 26 440 pounds of shellac were imported. In later years the quantity of shellac was twice that of stick lac, but considerable quantities of both were imported; in 1760, stick lac amounted to 170 780 pounds as compared with 349 630 pounds of shellac. Naturally, it was not all used as a dye, for lac had other uses in lacquers and varnishes.

During the seventeenth century, ready-prepared lake pigment was often distinguished by a place name. Florentine lake and lakes from Venice and Antwerp are mentioned quite frequently in literary sources, the Italian products being most highly recommended. Hilliard states that lake from Antwerp is quite good, and Gyles follows him, stating somewhat grudgingly that it is 'indifferently good'. It is noticeable that two of the three places mentioned were the most important ports in Europe during the sixteenth century, and it seems likely that their reputation for superior lakes rested on the fact that they monopolised trade and had first choice of the best raw materials. The tradition that the best lakes came from there lasted throughout the seventeenth century, but it would be wrong to assume that good lakes were not made elsewhere at that time. Norgate does not refer to European sources but speaks of India lake as being the best. The use of that name by Norgate is of some significance because the early version of his treatise was written when direct trade between England and India was being developed by the East India Company, and it suggests that the English were likely to have manufactured their own red lakes at that time. A total lack of documentary evidence concerning such manufacture is remarkable, especially as instructions for extracting the dye from cochineal and making red colours from other red dyes are relatively easy to find. It would be ridiculous to suppose that those who were able to make carmine from cochineal were unable to make lake from Indian lac, and one must assume that some specialised knowledge never found its way into print. On the other hand, it is possible that English lakes were not as good as some, for, as Italian lakes were praised during the seventeenth century, those of French manufacture were recommended during the eighteenth.

It was important to obtain a good-quality Indian lake pigment, because as an artists' colour it presented some difficulties. General opinion amongst painters in oils was that it required much grinding and that it took several days to dry, so long in fact that it necessitated the addition of an artificial drier. If stored in bladders, which were the usual containers for oil colours, the colour had a tendency to grow fat and unusable. There appears to have been some difference of opinion as to how much grinding lake required for preparation as a water colour. Nevertheless, there was general agreement that, in addition to the usual gum medium, a small addition of sugar candy was required to prevent cracking in the shell. Some sources contain the suggestion that a little ear wax should be mixed in as well, the reason, implied but not explained, being that the addition would improve the flow of the colour. It is evident that the actual paint presented some difficulties in use, whether it was oil or water colour, but because of its transparency and its colour (deep crimson, almost purple) it was extremely versatile. Indian lake could be used in full saturation to shade other reds, used on its own as in the crimson drapery backgrounds described by Norgate, and, above all, it could be reduced most successfully with a large proportion of white to provide excellent flesh tints. Its versatility made it well worth its high price, its value being second only to ultramarine.

Carmine

'Sanguine. Lake of spayne is most redd the finest for Carnation. Lake of Venice is most murrey and finest for glasing.'[239] Thus a writer of the early seventeenth century points out one of the differences between lac lake and cochineal; whereas lac lake from the East has a purple cast, lake from Spain (that is, from Latin American cochineal) is a truer red. This fundamental difference between the two pigments sometimes makes it possible to determine which lake is indicated in documentary sources written before the adoption of a special name for cochineal lake near the end of the seventeenth century. Cochineal is certainly meant in the passage quoted above, for the Spaniards had imported the dried bodies of cochineal insects, *Coccus cacti,* since development of American trade began in the sixteenth century. Other documentary sources may be less specific, but a description sometimes points to the use of cochineal, as in the passage below which is quoted from a letter written to Henry Gyles:[240]

> Dr Ashenden had a pound of Gum Lac which was English such as Vermuling useth. You may have indian lac at the same price; it is 20*d*. per pound very good. The English lacc is easier to work but the other will indur longer.

In this case there is no distinction about colour, but the description of the English lake as less permanent but easier to use than Indian lake is an accurate observation of two further differences between cochineal and lac lakes. The letter was written in 1682 at a time when the name *carmine* was occasionally used for cochineal lake (it had appeared in Browne's *Ars Pictoria* of 1669), although it seems possible that the name was not thoroughly naturalised until the end of the century. The use of the term English lake is most unusual and should be taken as an indication that the lake was of English manufacture, not that the raw material was of English origin.[241]

There are English references to the treatment of the dried insects for the production of a dye, and it seems that cochineal was imported in dried form and not given preliminary treatment in America in the way that lac was treated in India. The Spaniards are said to have prohibited the export of live insects from America, so that they could preserve a monopoly in cochineal.[242] However, the insects were also to be found in the West Indies, and there was no secret about the method of cultivation. A description of the insects and their collection from the cacti in which they live was published in English in 1693:[243]

> These Grubs in process of time becoming Flies, like our Lady-Birds, . . . and being come to full Maturity, (which must be found out by experience in collecting them at several seasons) they Kill by making a great smother of some combustible matter, to Windward of the shrubs whereon the Insects are feeding, (having before spread some Cloths all under the plants) whereby all the Insects being smother'd and Kil'd, by shaking the Plants will tumble down upon the Cloths. Thus they are gathered in great quantities with little trouble. Then they spread them on the same cloaths in some bare sandy place, or stone pavement, and expose them to the heat of the Sun, until they are dry, and their Bodies shrivel'd up.

Unlike lac insects, which are dead when collected, cochineal insects must be killed and dried. Exposure to the sun is still considered the best method, although hot water or oven heat may be used instead. In the past, both cochineal and the pigment made from it were extremely costly, and large numbers of insects had to be collected in order to make a small quantity of colour. A modern authority states that 50 000 weigh only two pounds and that 70 000 are required to make one pound of carmine.[244]

References to the manufacture of red lake pigments are scarce in seventeenth-century documents, and the few which are to be found allude to a traditional method whereby the dye was extracted from red cloth. Because cloth, usually torn up into rags, provided the colouring matter, the original source of the dye is not mentioned. One such recipe appears in an English manuscript which probably dates from the very beginning of the seventeenth century, although the recipe itself is likely to be earlier in origin:[245]

> To make Sinaper
> *Recipe* fflockes of scarlet one pound and one pinte bagg lie and put them togeather and let them stand 3 dayes and euer twiste or thrise stirre it well and then take it and grynde with a moller and yf the couller be to feeble put it in the lye againe a day and a night and then grynde it againe and then make it in small balles and let yt lye.

One of the methods explained in great detail by Neri in the early seventeenth century incorporates the traditional use of cloth, thereby making the whole process extremely devious. He specifies the use of cochineal, explaining how to extract the dye by means of tartar and then dye the cloth. He goes on to explain how to prepare an alkali in which to soak the cloth in order to extract the dye once again, and finally describes the preparation of alum water and explains how it should be added to the then coloured alkali to bring about the chemical reaction that produces a solid colour from the liquid. Neri's instructions were translated into English and printed in the second half of the seventeenth century, but, even though the statement that red lake is made from red shearings or rags is to be found in some eighteenth-century books, largely those compiled from earlier sources, the traditional method incorporating the use of cloth was descredited by the end of the seventeenth century. In the 1701 edition of John Smith's *The Art of Painting in Oyl* the author states that he cannot discover a method for making carmine; even so, he gives instructions for making a cochineal red:[246]

> Buy at the Drugists some good Cochinele, about half an ounce will go a great way. Take Thirty or Forty Grains, bruise them in a Gally-Pot to fine Pouder, then put to them as many Drops of the Tartar Lye as will just wet it, and make it give forth its Colour; and immediately add to it half a spoonful of Water, or more if the Colour be yet too deep, and you will have a delicate Purple Liquor or Tincture. Then take a bit of Allum, and with a Knife scrape very finely a very little of it into the Tincture, and this will take away the Purple Colour, and make it a delicate Crimson. Strain this through a fine Cloath into a clean Gally-Pot, and use it as soon as you can, for this is a Colour that always looks most Noble when soon made use of, for it will decay if it stand long.

A solution of tartar was normally used to extract the dye from cochineal; it is mentioned not only by Neri and Smith but also by Hoofnail, who discusses the manufacture of red lakes at some length in his *New Practical Improvements,* denouncing the traditional method of using red cloth on the ground that it produces a very dull colour. He outlines various methods and finally recommends that cochineal should be boiled in alum water, strained, and the liquid mixed with a solution of salt of tartar. He suggests that the addition of a little iron makes it more crimson (when in fact it is liable to make it black). Dossie likewise criticises traditional methods but adds nothing to information on the subject of making carmine, saying that he could not find a reliable process written in English or French.

Berger's records contain many formulae for lake-making; the following is a costing for a cochineal lake dated 2nd April 1787:[247]

6lb Cochineal boyled with 8oz Tarter and 9oz Tarter to Throw Down, 1 Tub White 21lb Ash and 42lb Allum, struck boyling Hott washed once hott and once Cold	5	12	6
2 Pails of old White		4	—
Fyer Labour etc	1	10	—
½lb Vermillion — 6/—		3	—
	£7	9s	6d

Vermilion may well have been included to improve the permanence of a very fugitive pigment; several eighteenth-century authors suggest that the lighter, scarlet varieties of carmine were adulterated.

Field states that a scarlet colour is made from cochineal by the agency of tin, and in Berger's later records it is indicated that alum, cream of tartar, borax and stannous chloride or stannous nitrate were used. That firm was producing purple lakes from cochineal as well as the red colour better known to artists.[248]

Carmine has been identified in an oil painting by Van Dyck, and, with improved methods for identification of the raw materials used in lake pigments, further examples may be found.[249] Bardwell's description of the pigment in the eighteenth century is likely to be representative of opinion at that time: 'It is a middle Colour between Lake and Vermillion; it is a fine working Colour, and glazes delightfully. It should be ground with Nut Oil and used with drying Oil.'[250] At the end of the eighteenth century the anonymous author of *A Compendium of Colors* stated that carmine was used less in oil than as a water colour on account of its high price, and it seems to have been recommended more frequently and certainly more extravagantly in books on water-colour painting, some authors (including Payne and Henderson) wrongly describing it as permanent. Although genuine carmine fades rapidly on exposure to light, it seems that a colour of better durability was sometimes produced by adding another pigment to it, as demonstrated in the method quoted from Berger's records. Very likely Ackermann's carmine lake was a mixture, as it was sold at two shillings per cake as opposed to five shillings for a cake of genuine carmine. Field states that a mixture of carmine and vermilion is sold as scarlet lake. Water-colour painters may have objected to such a combination, because the working properties of the two colours are very different and the presence of vermilion may have made it difficult to produce a smooth wash. Another variation of carmine, a purple sold under the name burnt carmine, was made by roasting genuine carmine until it reached a colour equivalent to that of gold purple. It is mentioned in books written during the late eighteenth and early nineteenth centuries, although the idea of roasting red lake is of older origin, as a burnt lake black is mentioned by de Mayerne.[251] Both burnt carmine and the more usual crimson colour shared the same outstanding characteristics of smooth, even flow which, despite their lack of permanence, made them very popular with painters in water colours.

Madder

The early history of madder red as an artists' colour is problematical because, whereas several varieties of the genus *Rubia,* the madder plant, are indigenous to Europe and the Near East and the roots are known to have been used for textile dyeing since early times, its use in painting is poorly documented. Writing of the medieval period, Thompson points out that madder may have been used alone to manufacture a pigment at

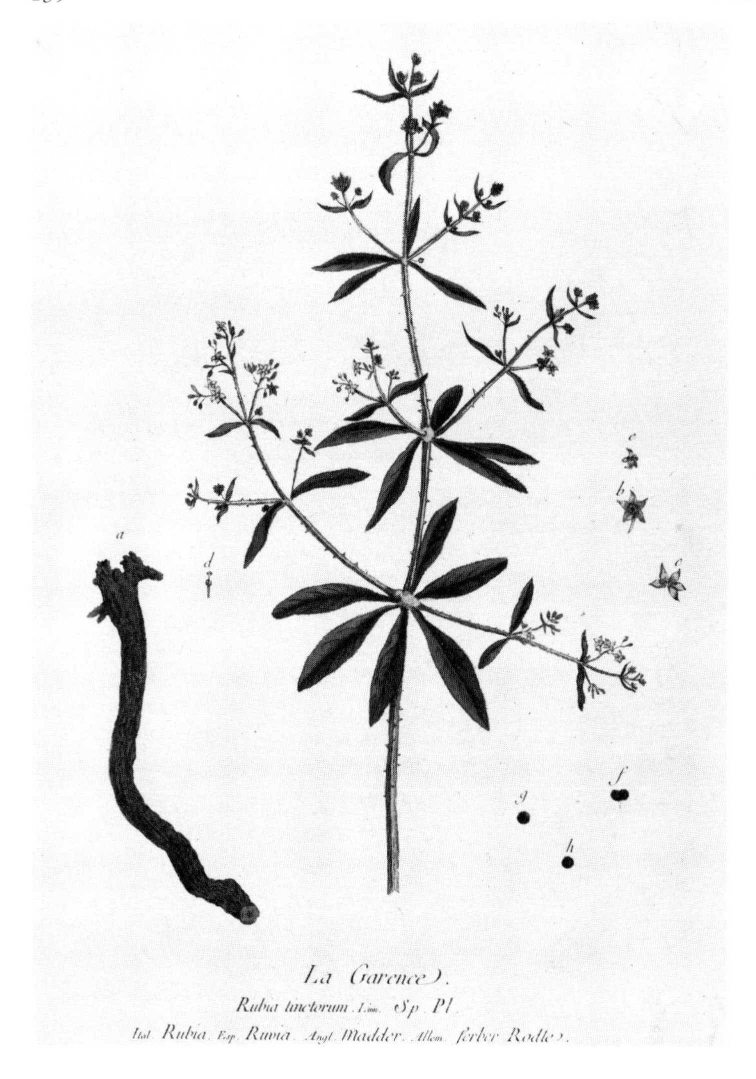

Fig. 42. Madder plant and root: from Regnault, *La botanique*, 1774. (Reproduced by courtesy of the Royal Horticultural Society.)

one time and then have been partially abandoned in favour of brasil wood, the two raw materials possibly being used together.[252] Documentary sources from the Middle Ages and the early modern period provide little information, largely because, as previously mentioned, colour-makers obtained dyes from cloth. This being so, it is likely that madder dye was employed in the manufacture of red lakes much more often than written evidence would suggest.

Seventeenth-century sources which are not concerned primarily with painting contain a number of references to madder plants. Gerard's herbal, which contains a somewhat diagrammatic illustration of *Rubia tinctorum,* also contains a description of the small, herbaceous plants with trailing stems which the author likened to Ladies Bedstraw. In his time the plants were common in many parts of England; they flourished from May to August; in the autumn the roots were gathered and they were 'sold to the use of Diers and Medicine'.[253] Madder plants grow wild and have also been cultivated, possibly not continuously but certainly continually since Roman times. During the Middle Ages the Moors cultivated madder in Turkey and also in Spain, both places where it grows still. The importance of cultivation in the Netherlands during modern times and the fact that by the seventeenth century the Dutch were the most advanced growers in Europe have tended to overshadow accounts of cultivation in other European countries, although the success of the French industry has sometimes been recorded.[254]

Anglo-Dutch rivalry was a formative influence in the history of the seventeenth century, as it affected events overseas and also British domestic policy. Madder was used so extensively in dyeing that the British became anxious concerning their reliance on the import of the Dutch product, with the result that royal approval was given to a madder cultivation and trading monopoly. In 1625 patent rights were granted for the purpose in the hope that the patentee would ultimately manage to grow and prepare enough for domestic requirements and produce a surplus for export.[255] The native industry was still in existence eleven years later, as a Dutchman who offered to pay an additional levy to the Crown above the usual customs dues for sole right to trade in madder was refused on the ground that the existing patent had nine years to run.[256] In the same year there was trouble concerning land which was covered by a proviso that madder must be grown there, and in 1637 an indenture was drawn up between the Crown and the Society of Planters of Madder of the City of Westminster (associates of the original patentee), agreeing upon dues to be paid to the Crown and the appointment of a government surveyor to supervise the management of the business, described as 'sowing, setting, planting, breaking, drying, dressing and preparing of madder'.[257]

The English industry was attempted long after the Dutch was well established, and Dutch influence is apparent in documents concerning the madder monopoly, many of which include special terms for madder of various qualities, all of which are of Dutch origin. In the indenture between the Crown and the madder growers, the best sort was described as *crop* and *unberoofde,* the second as *gemeene* and *fatt* madder, and the third and cheapest as *mull* madder. Of these terms only the first, crop madder, survived as a specialised but understood phrase in English meaning the best, inner part of the root. It has no connection with cultivation or the English word crop but is associated with the Dutch *meekrap* which means madder. The meaning of the other adjectives might have fallen into complete oblivion had it not been for Bancroft, who, in his book on dyes and dyeing published in the early nineteenth century, explained how madder roots are treated. After earth and dust had been shaken from them, the roots were dried and placed on wooden blocks and then stamped or pounded so that small roots and the husk of the larger ones were reduced to powder which was then sifted and sold as mull. A repetition of the process separated half the remaining roots, the powder once again being sifted and packed and then sold under the name gemeen. Final stamping reduced the inner part of the root (which is brighter in colour, orange as opposed to the brown of the outside) to a usable condition. Seventeenth-century records suggest that crop and unberoofde were of similar quality and sold at the same price, although, according to Bancroft, unberoofde was a mixture incorporating cheaper grades which yielded less dye than crop.[258]

Madder of all qualities was imported from Holland during the eighteenth century, and the specialised terms appear in the customs records. It is obvious that the English madder-growing society never succeeded in producing sufficient quantities during the seventeenth century, although national efforts to cultivate madder did not completely die out, for when the Society of Arts offered a subsidy of five pounds for every acre planted with madder, seventy-eight successful claims were made for the award.[259] However, domestic efforts were quite unequal to meet the huge demand which is reflected in import figures running into thousands of tons of prepared madder every year throughout the eighteenth century. A noticeable feature of the customs figures is a steadily growing increase in the import of untreated madder roots at that time. Crop madder always formed the greatest proportion of the total quantity, but, during the second half of the century, increasing quantities of roots were imported from Turkey. A modern authority states that Turkish madder was superior because it was sun-dried in the open instead of oven-dried as was normal practice for the European product.[260]

When discussing the manufacture of madder lake, little can be said about the seventeenth century apart from pointing out Merret's translation of Neri's instructions concerning both madder and brasil lakes. He recommends that less alum should be used in either than for cochineal lake and that a greater quantity of madder or brasil should be used to dye the cloth: 'And in this manner you shall have a very fair Lake for Painters, and with less charge than that from Cochineel, and that from Madder in particular will arise most fair and very slightly.'[261] Not until the early eighteenth century does one find extensive discussion on the subject in English, and, once again, it was Hoofnail who printed his opinion on the matter. His suggested method for making a good, reasonably cheap scarlet was to place two ounces of crop madder in one pint of alum water, heat it gently for two hours and allow it to stand for twenty-four, after which it was strained and precipitated with a solution of salt of tartar. It seems possible that Hoofnail encountered some difficulty, as a remark is included to the effect that the same quantity of madder does not regularly produce the same quantity of colour. A point Hoofnail did not mention, but which has been the subject of comment by later writers such as Thompson, is the fact that it is much more difficult to extract the red dye from madder than from other raw materials such as brasil wood.

There is little evidence concerning the commercial use of madder in the manufacture of artists' colours until the beginning of the nineteenth century, although it seems likely that it was used during the eighteenth century, for the colour-maker de Massoul states that the lake least given to colour alteration is made from madder. It is impossible to tell from his list which colour incorporated madder because it is nowhere mentioned by name, and the conclusion must be that it was simply called lake. In 1804 the Society of Arts published an account of their award of a gold medal to Sir Henry Englefield for the development of a lake from madder having the depth and transparency of carmine, unlike existing madder lakes which were said to be too yellow, too pale or a brown-red. The lake-making method is outlined with suggested variations, and an interesting feature is the comment that the colouring matter of madder is not easily soluble in water, so necessitating the use of mechanical pressure to assist the operation.[262] This point may have been observed at an earlier period; in fact it would be surprising if dyers who had worked with madder for centuries had not arrived at a good method of extracting the dye, but the observation does not appear to have been made in books on colours or painting.

At the beginning of the nineteenth century, however, the necessity for extracting the best, red colouring matter from madder under pressure was fully realised, and, from an illustrated account of Field's apparatus, it is obvious that he used a press for extracting the dye, the madder being held in a bag which was under continuous pressure from a weight and screw at the top, while water ran through the bag and its contents, carrying the dye down to pour into a receptacle placed below. An account of the press and other equipment was submitted by Field to the Society of Arts, and in 1815 several members of the committee concerned with chemistry and mechanics made a trip to Hounslow Heath to see it in action. They decided that the press did not incorporate a new idea or principle, but Field was awarded a gold medal for the novelty of his physeter or percolator, which was designed to filter lake colours under pressure by pumping the liquid colour through, the pigment being collected by a sieve of copper wire covered by woollen baize and a silk cloth, while the unwanted liquid was pumped away.[263] This was a considerable improvement on the traditional method of filtering, because, according to all seventeenth-century and eighteenth-century accounts of lake-making, after precipitation had taken place, the colour was filtered through a cloth or conical bag so that the solid colour remained in the cloth while the liquid dripped through. Afterwards the colour was removed from the bag, broken into lumps and dried on bricks or pieces of chalk, sometimes in the sun, ready for crushing and powdering into pigment. Some progress in mechanisation had obviously been made by the nineteenth century, because, just as Field's dye extracting press was judged to be a variation of an existing idea, his drying oven was not new. Nevertheless, both pieces of equipment were illustrated with traditional, conical bag filters and Field's award-winning percolator or pressure filter in the *Transactions* of the Society of Arts.[264] All Field's devices imparted not only increased efficiency but also greater speed in the completion of a lengthy operation, both important considerations in any commercial undertaking. Field did not publish any comments on the chemistry of lake-making, although relevant comments are to be found in his manuscript notebooks, as also are sketches of his factory at different periods (see *Figures 52* and *53*). Detailed comments on the subject from much the same period are to be found in Mérimée's book, which, although it is concerned mainly with French manufacture, mentions the use of some equipment similar to Field's.[265]

English lake-making did not achieve a sound reputation until the nineteenth century, and its success then appears to have been largely owing to the efforts of Field. In his obituary, he is said to have turned his attention to madder when supplies were difficult to obtain during the Napoleonic wars.[266] The statement is not supported by the customs records, which show that supplies were obtained from Holland, Turkey and Italy, and also from France in foreign ships. Field himself claimed to have already been engaged in the commercial production of madder lake before Englefield gained the Society of Arts award in 1804, although he did not explain any particular reason for starting to do so, apart from the fact that he was encouraged by the portrait painter George Joseph.[267] By the time *Chromatography* was published Field manufactured a range of madder pigments comprising rose, brown and purple shades and a red called madder carmine which was superior to rose madder in texture and transparency. In addition, he prepared madder in the form of a concentrated liquid for water-colour painting under the name 'liquid rubiate'. Samples of Field's rose, brown and purple madder are to be found in his 'Practical Journal 1809', folios 312 and 313. (See *Plate 1*.)

Madder as a colour name came into use in the early nineteenth century. Rose, purple and brown madder are all mentioned in a manuscript treatise on water colour by John Thirtle, an artist of the Norwich school.[268] In Hamilton's *The Elements of Drawing,* madder is not included in the long colour list although rosy madder appears elsewhere in the book in a list of colours used by Varley. The information is doubtless correct, as, although there is no sample of the colour in *J. Varley's List of Colours* of 1816, madder lakes, including brown and purple, are mentioned amongst useful colours in the short notes following the samples. Dagley lists both lake and madder lake, and Fielding includes washes of both, mentioning that madder lake is a little difficult to apply successfully. This point is also touched on by Field, who makes an exception for his own madder carmine and adds the information that rose madders in oil are fairly slow driers.

Purple madder is not included amongst any of the printed books of the period containing colour samples, although there are examples in Field's manuscript notebooks. The colour that Field made and sold under the name purple rubiate or Field's purple was described by him as rich and deep without brilliance. It was more useful than either gold purple or burnt carmine on account of its transparency, which is a feature of all lake pigments, and its comparatively good durability, that is shared by the lake pigments made from madder.

Brasil

Brasil is a comprehensive term for the hard, brown-red wood obtained from trees of the genus *Caesalpinia,* varieties of which are found in the East Indies and South America, and for the dye obtained from the wood. The name itself has given rise to misunderstanding, for some writers of the eighteenth and early nineteenth centuries were under the impression that the country called Brazil gave its name to the wood, although the reverse was actually the case, since the best brasil wood capable of yielding a large quantity of dye was discovered near Pernambuco and was for many years exported from there. The word *brasil* originally meant red; it developed from the same root as the Latin *rosa,* as also did the English name *roset* which was at one time used for the artists' colour made from brasil dye. The similarity, or possibly confusion, between brasil and roset can be seen in the use of the names *roosyle* and *rosselte* in two English manuscripts.[269]

At first sight it seems strange to find specific references to brasil to the complete exclusion of madder in early sources, and also to find the colour made from brasil being distinguished by a special name. It is noticeable, however, that the name was not used for a lake pigment made from brasil, and, for that reason, one may assume that when brasil was used for lake-making, as it may well have been when red rags were

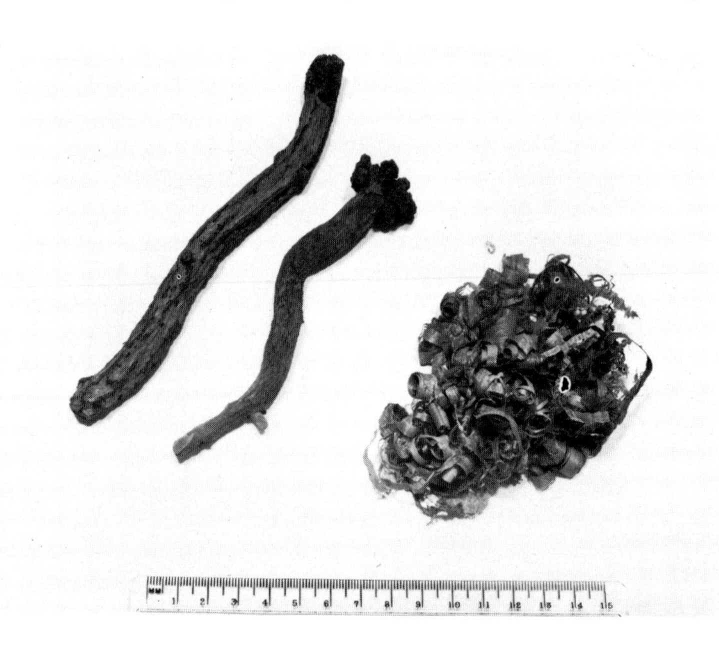

Fig. 43. Madder root, *left*, and brasil wood raspings, *right*, both raw materials from which red dye may be obtained. (Reproduced by the kind permission of Messrs Winsor and Newton Limited.)

employed for the purpose, it remained just as anonymous as madder. The reason why brasil was featured in early sources rests on the fact that it was relatively easy for the painter to extract the dye from brasil and to prepare it in the form of a concentrated liquid for water-colour painting.

Brasil wood was imported in blocks which were then rasped and reduced to a coarse powder which in the early seventeenth century was sold at tenpence per pound, much more cheaply than cochineal which is listed at thirty-five shillings in the same manuscript.[270] By the late eighteenth century the price had altered very little, as brasil raspings were available from dry-salters at one shilling per pound. Colour-makers or painters prepared brasil by soaking and boiling the powder with water which was often mixed with vinegar. Alum was also added, and the resulting colour was usually tempered with gum solution. A number of written instructions imply but do not really explain that a concentrated form of colour is obtained by boiling the dye. Both de Mayerne and Goeree give detailed instructions for making the colour. A translation of de Mayerne's process is as follows:[271]

> Excellent Roset. Boil three ounces of good, Pernambuco brasil of a strong, dark red together with one ounce of alum in a quart of strong red wine vinegar until half the quantity has evaporated. Pour it out and add half an ounce of good, clean gum arabic which makes a fine *colombine.* One ounce of gum makes it deeper; it must be added at the end of the process. Pour out your roset. It keeps well for several years and needs no precipitation.

Goeree's instructions are similar, although the quantities are larger; half a pound of brasil to half a pint of stale beer, a glass of water and a similar quantity of vinegar, one and a half ounces of alum and a half ounce of gum tragacanth are recommended. He likewise states that half the quantity must be boiled away and that the resulting colour will keep well: 'the older this is, the better will the colour be.'[272]

An interesting feature of de Mayerne's recipe is the inclusion of the phrase 'qui faict vn beau columbin'. The French word *colombine,* which is used in textile dyeing, means dove-coloured or the colour on the throat of a dove. It seems to have been applied particularly to brasil colour or dye, because in Boutet's book on miniature painting *lacque colombine* is the name used for a pigment made from brasil. The French name is relevant to a discussion of English names purely because Boutet's book was translated into English in the first half of the eighteenth century and the English name *rose pink* made its first appearance in that translation. By that time the name *roset* had become obsolete, for when a liquid water colour made from brasil is mentioned in eighteenth-century books it is always called simply brasil, but there appears to have been a desire to distinguish between brasil liquid and the pigment, and, consequently, *lacque colombine* was translated as rose pink. Obviously the adjective which gave an indication of the colour was rose. Pink was almost certainly used as a noun, standing for an association of ideas between the rose coloured pigment and the green-yellow colour called pink which was made in the same way, being a pseudo-lake pigment. Dossie describes brasil lake as 'rose lake commonly called rose pink', but descriptions of rose pink point to the fact that it always contained chalk (as also did English pink) and that it was an inferior colour. Consequently, it was not a true lake made by means of a chemical reaction, but a less brilliant pigment of inferior durability made by mordanting brasil dye on chalk.[273]

The pigment, rose pink, never became popular during the eighteenth century. Dossie dismisses it as suitable only for coarse work, and Field, writing in the nineteenth century, states that, although it may be used by paper stainers, it is unsuitable for artists. For artistic purposes brasil was always more important as a liquid for water-colour painting than for oils, and even the liquid was not very important after the mid-seventeenth century. Until that time it was mentioned frequently in books on limning and washing, sometimes described as crimson, as in B.M. MSS. Sloane

3292 and Stowe 680, and sometimes as having a brown cast. Brasil is not light-fast. Whereas it may have been suitable for limning in books which remained closed much of the time, the colour was not good enough for portrait miniatures which were likely to be exposed to light and it was excluded from Norgate's palette for that reason. After his time, brasil red was recommended only for washing prints and maps on the ground that such painting, practised by amateurs, was not expected to be of a durable nature. In modern terms, lac lake, carmine and, later, madder were artists' quality colours while brasil was included only in the economy range.

Dragon's Blood

This name is applied not to a lake pigment but to a dark red resinous substance that is obtained from certain trees indigenous principally to eastern Asia, the variety usually named being *Calamus draco.* It was known in ancient times, and the fact that the dry resin looks similar to dried blood probably gave rise to the legend that the substance was mingled blood of dragon and elephant. Dragon's blood was not much used in painting during the seventeenth century, for it is mentioned only occasionally in documentary sources, usually appearing under its Latin name *sanguis draconis.* It is listed without description in B.M. MS. Sloane 6284, f. 109v, and it is included in the price lists in MSS. Stowe 680 and V. & A. 86.EE.69. Although it is not listed with other colours in B.M. MS. Harley 6000, Norgate's treatise contains a suggestion that it may be used for superimposing on silver or other metals: 'Dissolue Sanguis Draconis in aquavitie on the ffyer, which done with a pencell lay it on and lett it drye.'[274] A suggestion that it should be ground with red wine vinegar and used to shade carnation, that is, flesh colour, is included in MS. Sloane 3292, f. 4. It is obvious from the last two references that the transparency of dragon's blood made it useful for glazing purposes. Before the colour could be used, the resin had to be dissolved, for it was purchased in dry pieces. De Massoul refers to the pieces as red drops and adds that they are imported enveloped in leaves. Field's comments adjoining the colour sample of dragon's blood in his 'Practical Journal 1809', f. 326, include the fact that the raw material was in sticks and that in oil the colour dried with difficulty, it was extremely transparent, and, in thin applications, browned and faded rapidly. It seems unlikely that dragon's blood was much used during the eighteenth and nineteenth centuries, because relatively small quantities were imported, the annual total seldom exceeding one thousand pounds. The few references to it in literary sources of that period suggest that the resin was used more as a colour in varnishes than in paintings.

Safflower

Safflower is a common name for *Carthamus tinctorius,* a plant which grows in the warm climate of southern Europe and Asia. The red dye obtained from the dried petals of the flowers was at one time in demand for dyeing government red tape, and it was also used for making an artists' pigment in the nineteenth century. Unfortunately, safflower and saffron, *Crocus sativus,* are commonly confused, and it is reasonable to suppose that some nineteenth-century writers meant safflower when they mentioned saffron. In Alston's book a rose colour is described as being made from saffron: 'Saffron flowers prepared in saucers are sold in some shops, called saucer colour.'[275] This is followed by the information that at one time it was obtainable only from French milliners but by 1804 it was more common and could be bought at colour shops. In Towne's book, published in 1811, a red colour is mentioned by the name *assiette rouge,* which is comparable with the English name *saucer colour.* Saffron *en godet* is mentioned in *Le maître de miniature,* 1820. A suspicion that all the colours mentioned were made from safflower is confirmed by reference to Field, who lists the names rouge vegetale, Chinese *rouge,* and pink saucers, saying that the colour sold under those names is prepared from safflower. He adds the information that it is used in dyeing silks and for cosmetic purposes and that, although beautiful and costly, it is too fugitive for the use of artists. Judging by the absence of earlier references, it does not seem to have been used before the nineteenth century. Turner may have been one of the painters who used it at that time, as a deep red powder amongst pigments which once belonged to him has been identified as an organic red, possibly safflower, on a base of alumina.[276]

11

Browns, Blacks, Greys

Brown Ochre and Umber: Fe_2O_3, MnO_2, clays, etc.

Brown ochre has already been mentioned in connection with yellow and red ochres, for there can be no firm division between brownish yellow or red oxides. Ochre is most often listed as yellow in literary sources, although *brown ocker* is listed with other oil colours in V. & A. MS. 86.EE.69 and *bole armoniake,* generally regarded as red, is listed with browns in Bodleian MS. Ashmole 1494. A brown called Nottingham colour, presumably an ochre, is listed with browns by Henry Gyles; it was used as a water colour for shading yellows. However, the brown ochre mentioned most frequently is the colour known as Spanish brown, which in the seventeenth century and subsequently was prepared in England by heating native red ochre. Brown ochre was not an altogether satisfactory colour. Dossie, writing in the eighteenth century, states that it is prepared cheaply in large quantities for coarse work. Similarly, the pigment prepared commercially in the seventeenth century was more suitable for decorating than for artistic painting, as Norgate condemned it as 'exceeding course and full of gravell, and of noe greate use since umber and a little Lake tempered togeither make the same Collour'.[277]

In comparison with brown ochres, umber is mentioned much more frequently. This brown earth bears some similarity to ochres, but, in addition to iron oxide, it contains manganese dioxide which imparts slightly different characteristics. When discussing umber some past writers comment on the colour name and a few, including the anonymous author of the *Practical Treatise,* 1795, state that the pigment is called umber after the district of Umbria in Italy, whereas others, such, as the French writer Watin, suggest that the name means 'shadow' because it is derived from the Latin *ombra.* The latter suggestion seems very likely, as umber is often described in documentary sources as being very suitable for shadow or for shading other colours, and, although the earth is found in a number of places in Europe, north Africa and the Near East, most umber was imported from Turkey, not from Italy. The statement that most umber is imported from Turkey, which is made in the *Practical Treatise,* is borne out by the customs records that show that several tons of umber were imported from Turkey in as many years whereas the raw material is not listed as coming from any other country. Some of the best umber is reputed to come from Cyprus, which would have been classed as part of Turkey in the eighteenth-century customs records.

Umber can be used in its raw state or after it has been roasted, when it has a deeper brown-red colour. The distinction between raw umber and burnt umber is often ignored in documentary sources of the late sixteenth and early seventeenth centuries; for example, it is listed merely as *umbra* in V. & A. MS. 86.EE.69. A little later, however, writers, including de Mayerne and Gyles, make the distinction, and umber burnt and unburnt are also listed in *Academia Italica.* It is possible that burnt umber is meant in any place where it is listed without further description, for burnt umber was of greater use and easier to work with according to Norgate. Umber appears in his colour list without qualification, but a subsequent description makes it clear that the artist used only burnt umber. The pigment is described as follows:[278]

> a Collour greasy and foule, and harde to worke withall if yow grinde him as hee is bought, yett of very greate vse for Shadowes and hayres etc. Yow must burne it in a Crusible or Gould smithes pott, by which meanes it is clensed, and being ground as the rest works sharpe and well.

De Mayerne explains that umber must be roasted until it reddens, when it is equal in colour to a mixture of raw umber and lake but is much more durable.

An outstanding feature of umber in oil is its quick-drying property, which is imparted by the manganese content. De Mayerne remarks that it dries in two hours, and Marshall Smith suggests that after it has been ground in linseed oil it should be put in a gallipot and kept under water so that it shall not dry before it is used. A number of seventeenth-century writers recommend that umber should be used as a drier in conjunction with lamp black. It is very difficult to understand Dossie's comment that umber was used a lot at one time but was neglected in the mid-eighteenth century, for it is recommended as a good, useful colour for painting in oils and water colours, not only in seventeenth-century treatises but also in eighteenth-century works. Its permanence and translucency make it suitable for oils and for all types of water-colour painting. Umber is recommended by Hilliard for shading browns and blacks, and in *Academia Italica* for colouring maps and prints. Many writers indicate that it is an essential colour, for it is marked by an asterisk in Williams' list of oil colours and it finds a place in the restricted palettes of late eighteenth-century and early nineteenth-century writers on water-colour painting. General opinion seems to be that it was a dependable colour which was useful for many purposes. Field, who was particularly interested in durability, suggests that umber may with advantage be used instead of Vandyke brown.

Vandyke Brown (Cassel Earth, Cologne Earth)

The present-day name of this dark brown transparent colour appears to be relatively modern, being listed in late eighteenth-century books, such as Ibbetson's *Process of Tinted Drawing,* and those by de Massoul and Payne. The last named and Field are the only writers to explain that Cassel earth is named Vandyke brown because of the high esteem with which the seventeenth-century painter regarded the colour. The spelling of the colour name is generally anglicised, although a few past writers, such as Dayes and Roberts, use the alternative spelling Van Dyck. The late appearance of the name *Vandyke brown* may have given rise to an idea, put forward by modern writers, that such organic brown colours did not come into use until the late seventeenth century.[279] However, they are mentioned, usually under the name *Cologne earth*, in English sources dating from the beginning of the seventeenth century. Early examples are *earth of Cullen,* which is listed with blacks in B.M. MS. Sloane 6284, f. 109v, and is also mentioned by Hilliard as a colour suitable for shading browns and blacks. De Mayerne describes it as a red-black and names it in German as *Colniche erden.* Norgate gives it an Italian name and describes it in English: 'Terra di Colonia is easy to worke when it is new ground and is very good to close upp the last and deepest touches in the Shaddowe places of pictures by the life, and likewise very usefull in Landscape.'[280] Although it is not listed in all later books, it appears at intervals throughout the seventeenth and eighteenth centuries. Marshall Smith describes how *Colens earth* must be ground in linseed oil and then put out on brown paper to drain, after which it may be used with drying oil. The pigment was called an earth because it was dug out of the ground, but it is organic since it is derived from deposits of lignite or peat, and it is of a bituminous nature which means that it is a slow drier in oil, hence Smith's stipulation that it should be used with drying oil. Some eighteenth-century writers did not realise its true nature, for Watin states that it is a more transparent kind of umber. De Massoul repeats the statement but also remarks that it is a species of peat.

By the end of the eighteenth century two varieties were frequently mentioned; a distinction was made between Cologne earth and Cassel earth, Vandyke brown being synonymous with the latter. A colour sample of Vandyke brown is included in Ibbetson's *Process of Tinted Drawing,* and samples of both Vandyke brown and Cologne earth are supplied in Roberts' *Introductory Lessons.* The former is the darker of the two in one of the British Museum copies of Roberts' book. However, different batches of pigment of the same name may have varied, because Field states that Cologne earth is slightly darker than Cassel earth, the latter being more inclined to russet than the Rubens' brown used in the Netherlands, which is more of an ochre than the Vandyke brown used in England. According to Payne, Cassel earth or Vandyke brown could be coarse and sandy which made it troublesome to prepare, a process which he does not describe. Field states that its preparation consists of purification by grinding and washing over. He regarded the pigment as durable in both oil and water-colour painting, but modern Vandyke brown is classed as fugitive in both techniques.

Asphaltum

The bituminous raw material, which is found in natural deposits, is a mixture of hydrocarbons and is, therefore, organic, as is Vandyke brown.

Two important sources of asphaltum are the Dead Sea region and the island of Trinidad. The former has been known as a source since ancient times, and the occurrence of asphaltum in America was known in the seventeenth century and probably earlier. Parkinson's herbal includes an account of Lake Asphaltites, the best-known source in the Near East, and the information that 'ther are other sorts of Bitumen in the world . . . as in Cuba, and sundry fountaines neare the sea shore, casting it forth as blacke as Pitch'. Peru is also mentioned and the product is described as 'of a strong smell and of a blackish red colour. The Inhabitants about this Lake, gather this Bitumen or Pitch, being an oyle or liquid substance on the water and hardned by the aire, and spend it chiefly in pitching their Ships.'[281] This is presumably not a first-hand account, but it supplies the information that asphaltum is semi-solid when collected and that it is insoluble or barely soluble in water, thus making it suitable for protective coating on ships and other outdoor objects.

In view of the incompatibility between asphaltum and water, it is a little surprising to find that the pigment was used as a water colour, being mentioned in two sources *circa* 1600. In MS. Sloane 6284, f. 109v, *spalte* is listed with blacks, and *spalte* is likewise mentioned by Hilliard as a shading colour. After that time its use in water-colour painting appears to have been discontinued until asphaltum water colour was sold in bottles in the nineteenth century.

The colour, which is soluble in turpentine, was far more important in oil painting, and several books contain instructions for its preparation. De Mayerne specifies that when selecting spalte or aspalathum one must choose those pieces which are pure, very black and crumbly, and he gives the following instructions for its preparation which he obtained from van Somer. Asphaltum is not ground at all; instead a drying oil is made with litharge, then pulverised asphaltum and the oil are placed together in a glass vessel and suspended over the fire so that the contents melt like butter. It must be removed immediately it begins to boil. The resulting colour is fine for shading, glazes in the same way as lake and does not fade at all.[282] A late eighteenth-century formula for the preparation of asphaltum consists of simmering two ounces of copaiba balsam over a slow fire, adding one ounce of crushed asphaltum and as much turpentine as is necessary to dissolve it, the whole being heated cautiously so that it does not catch fire.[283] In the *Practical Treatise* the method is attributed to the painter Richard Wilson, perhaps wrongly, because a modern writer has pointed out that of all Wilson's paintings only one early example shows the cracking or 'alligatoring' characteristically associated with the use of asphaltum.[284] In *A Compendium of Colors* virtually the same formula is attributed to Mengs.

The oil colour is generally described as asphaltum, but the name Antwerp brown is applied to it in Williams' book. The author states that, in order to make it, asphaltum should be placed in a ladle over the fire, and, when it no longer boils, half an ounce of sugar of lead should be added to an equal quantity of asphaltum and both should be ground together with very strong drying oil. Williams states that Antwerp brown works freely and dries well, and Field, who also describes it as a preparation of asphaltum in drying oil, remarks that it is less liable to crack than an unground preparation of asphaltum.

In both the seventeenth and eighteenth centuries asphaltum was recommended as a colour for glazing and inserting shadows in the final stages of a painting; flesh shadows are mentioned specifically in almost every case. The colour's popularity during the second half of the eighteenth century led to indiscriminate use, and the poor condition of some paintings from that period may be attributed to the presence of excessive quantities of asphaltum. Despite all attempts to make the colour dry properly by preparing it with a metallic drier, it often failed to do so. Asphaltum has achieved notoriety on that account, but some writers were aware of the problems associated with the colour even when it was most popular. For example, the author of *A Compendium of Colors* states that asphaltum is used too much. Some years later, Field was much more outspoken, saying that its perfect transparency led artists to use it 'notwithstanding the certain destruction which awaits the work on which it is much employed'. His words may have had some effect in dissuading painters from using asphaltum, but it seems equally likely that fashion, not technical considerations, dictate whether or not a painter used the colour. Asphaltum was extremely fashionable amongst those eighteenth-century painters who favoured *chiaroscuro* and deep, transparent shadows, and nineteenth-century painters who adopted a similar style continued to use the colour. Wilkie may be cited as an example.[285] On the other hand, the Pre-Raphaelites preferred more luminous pictures painted with brighter colours (many of the pigments for which were supplied by Field), and, in consequence, asphaltum, which had little place in their style or technique, began to go out of fashion.

Mummy

Mummified bodies imported from Egypt were used medicinally during the sixteenth and seventeenth centuries. Pieces of mummy were made up into a drug which was taken internally, and, as was the case with many drugs, the raw material was also tried out as a pigment. Judging by the infrequent references to mummy in documentary sources, it does not appear to have been used to a great extent in painting, but that may have been a result of limited availability rather than squeamishness on the part of painters.

The mummies that provided the material for the pigment did not come from the most important royal tombs of ancient Egypt but from large communal tombs in the same area as the Pyramids. During the sixteenth century, export of mummies from Egypt was illegal, but Europeans acquired a taste for them, and, as a contemporary traveller stated, 'with words and money the Moors will be entreated to anything'. In 1586 some English travellers were delighted with their visit to the *momia,* a sandy underground cave which contained a large number of mummies. Slightly apprehensive, they were let down by rope, walked about on the mummies and viewed them by candlelight:[286]

> They gave no noysome smell at all, but ar like pitch, beinge broken; for I broke of all parts of the bodies to see howe the flesh was turned to drugge, and brought home divers heads, hands, arms, and feete for a shewe. We bought allso 600 lb. for the Turkie Company in peces, and brought into Ingland in the *Hercules,* together with a whole bodie. They are lapped in 100 doble of cloth, which rotton and pillinge of, you may see the skinne, flesh, fingers, and nayles firme, onelie altered blacke.

Mummy was black in appearance but a transparent brown when used as an artists' colour, its properties being somewhat similar to asphaltum. Haydocke lists *mummia* as a shading colour for flesh tones, and de Mayerne also recommends it as a shading colour, stating that some white copperas must be added to it as a drier. Browne includes the colour in the general list at the beginning of his book but excludes it from the list in the appendix, having come to the conclusion that it was unsuitable as a water colour: 'Mummy is every way ill-condition'd, and hard, and will not flow out of your Pencil, unless you burn it in a Crucible well Luted; so prepared, it may make a good Black.'[287]

Further remarks about mummy are not to be found until the end of the eighteenth century, when presumably it suited the style of many oil painters in the same way as asphaltum. Instructions for its preparation are included in *A Compendium of Colors:*[288]

> The finest brown used by Mr West in glazing is the flesh of mummy; the most fleshy are the best parts;... it must be ground up with nut oil very fine, and may be mixed for glazing with ultramarine, lake, blue, or any other glazing colours; when it is used, a little drying oil must be mixed with the varnish, without which it will be longer in drying, which is the only defect it has, as it may be used in any part of a picture without fear of its changing.

A colour sample is included in Field's 'Practical Journal 1809' with the following entry:[289]

> *Mummy,* Egyptian, from Sir William Beechey, in a mass, containing and permeating rib-bone etc.—of a strong smell resembling Garlic and Ammonia—grinds easily—works rather pasty—unaffected by damp and foul air—somewhat weakened by long continuance in the Sun.

Mummy, or Egyptian brown as it was sometimes called, was more durable and less liable to crack than asphaltum, and, for that reason, Field suggested in *Chromatography* that it could usefully be employed as a substitute for that colour.

Prussian brown is sometimes mentioned in conjunction with mummy in late eighteenth-century books. The colour was prepared either by roasting Prussian blue or treating the blue pigment with an alkali so that it turned brown. It is highly praised in *A Compendium of Colors.*[290]

> The finest brown next to that of mummy is the Prussian blue burned, which is to be used in the same manner for glazing as the former, with

> this difference, from its being a better drier, there is no occasion to use drying oil with the varnish. In some respects, this has the advantage of the mummy, being very little, if at all, inferior in point of color; it dries better, is obtained with less difficulty, and ground with greater ease.

In spite of this good account of Prussian brown, few references to the colour are to be found, and one must assume that it was used less than mummy.

Bistre

A brown pigment prepared from wood soot may have been used since very early times, but it is not specifically mentioned in English sources of the sixteenth and seventeenth centuries. Hilliard's reference to soot for shading brown or black gives no indication as to whether wood or coal soot is meant, and the latter seems most likely. The French name *bistre* is listed in both of de Mayerne's manuscripts and in Boutet's book on miniature painting. In the English translation of Goeree's book the colour is listed simply as *soot of wood.* None of the four works listed is purely English in origin, and it is not until one reaches English sources dating from the second half of the eighteenth century that frequent references to bistre are to be found. It seems that the best pigment was made from beechwood in France, and Dossie attributes its limited use in England to that fact. Once the pigment was obtained, preparation was a comparatively simple matter. It was boiled in water, and, when it had been allowed to settle to some extent but was still hot, the clearer part was separated from the sediment and the fluid was evaporated, leaving a fine pigment of a warm, deep, transparent brown.

Bistre was used only in water-colour painting, for asphaltum provided a similar transparent brown in oils. Bistre may have been superior to asphaltum as a water colour, although it was not altogether easy to use, its slightly resinous character making it difficult to work with and mix with other colours. Spanish liquorice was often incorporated with it to improve its working properties and impart a richer tone, but such an addition was detrimental because liquorice brown was fugitive. In praising Ackermann's bistre, an early nineteenth-century writer points out some of the deficiencies of the pigment as normally supplied.[291]

> Bistre is one of the most useful colours that is made use of with water, but it is seldom met with, possessing all the qualities which it ought to have. There has lately, however, a new bistre been discovered by Mr Ackermann of the Strand, that is of the first degree of excellence, as it approaches to sepia, and will mix well with other colours, which the common bistre will not; its tint also, is of a beautiful hue, works with much freedom, and bears washing over, which latter quality common bistre does not possess.

The comparison of Ackermann's bistre with sepia is interesting, because when the latter was available in the nineteenth century it was much preferred to the pigment made from wood soot and so took its place. Consequently, bistre was little used in England except for a comparatively short period in the eighteenth century.

Sepia

The transparent dark brown called sepia was named after the cuttlefish from which it was obtained, most of the colour being prepared from the variety *Sepia officinalis,* numbers of which are found in the Adriatic Sea. Cuttlefish secrete a black fluid which can be released into the water as a defensive measure, and the fluid-producing glands were used in the preparation of the artists' pigment. According to a nineteenth-century account of the colour, it was imperative to remove and dry the ink sac or

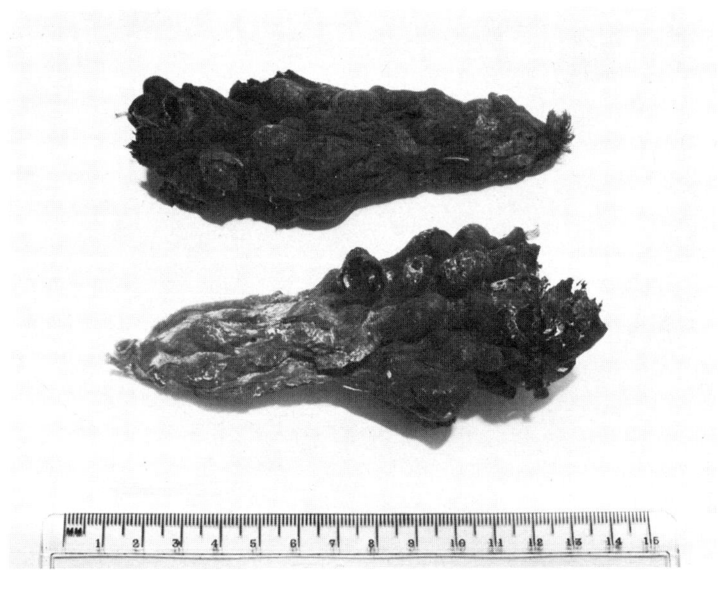

Fig. 44. Ink sacs from cuttlefish were extracted, tied together in bundles (two of which are illustrated here) and hung up to dry, after which they were used as the raw material for sepia. (Reproduced by the kind permission of Messrs Winsor and Newton Limited.)

gland as soon as possible after the fish was caught, as it putrefied quickly. The dried substance was first ground alone and then ground and boiled with an alkali of a caustic nature, filtered and neutralised with an acid. The brown precipitate was separated, washed in water and dried, the resulting pigment being a brown of a very fine grain.[292] From this account, which was written by Ure, it seems likely that the pigment was prepared in England, but pigment as well as the raw material was probably imported from Italy, as some nineteenth-century colour lists include both sepia and Roman sepia.

The fine grain and transparency of sepia made it particularly suitable for use as a water colour. Its value is indicated by its inclusion in most early nineteenth-century books on water-colour painting, many of which contain restricted colour lists which include only the most useful colours. Sepia was used in finished water-colour paintings, although

Varley suggests that it is most useful for sketches. He states that it is principally used in place of Indian ink owing to its richer tone. Sepia did not replace Indian ink entirely, but references to that colour and bistre are to be found a little less frequently after 1800 than in previous years when sepia was not available. A sample wash of sepia is to be found in Fielding's *Mixed Tints* and in Field's 'Practical Journal 1809', f. 338 (see *Plate 2*). The latter was obtained from the colourman, Newman, who had it in lump form at eight shillings per ounce.

Miscellaneous Browns

Some colours have been mentioned elsewhere even though, strictly speaking, they may be described as brown. They include the purple brown iron oxide known as Mars brown, which is mentioned in the chapter on inorganic reds, and dragon's blood, which is discussed with organic reds but is to be found listed with browns in B.M. MS. Stowe 680. Brown pink has been discussed in connection with organic yellows, although it was actually a yellow brown. The pigment was probably little used in oil towards the end of the eighteenth century, as some writers on oil painting make a point of mentioning that asphaltum mixed with a little yellow lake is superior to brown pink. Madder has been discussed in the chapter on organic reds. During the eighteenth century madder reds were criticised because they were too brown, but during the nineteenth century the colour deliberately prepared as madder brown or russet rubiate was extremely popular in water-colour painting. It is of course mentioned by Field, who specialised in making it, and it is mentioned also in the works of Varley, Dagley, Nicholson and Fielding, the last of which contains a colour sample.

Reference has been made to Spanish liquorice, because it was sometimes added to bistre. Dossie explains that the colour is made by extracting the succulent part of liquorice roots by means of water and evaporating it to dryness. He states that it is so sticky and gummy that it does not require the addition of any gum medium for water-colour painting. It is mentioned in a few books written during the second half of the eighteenth century, just as another organic brown, *terra japonica* or Japan earth, is also mentioned as a water colour in the same books. *Terra japonica* is not an earth but an extract, sometimes known as pale catechu, of East Indian plants of the genus *Uncaria.*

Manganese brown is an inorganic pigment mentioned by Field, who describes it as a fine, deep, semi-opaque pigment prepared from oxide of manganese. It is listed by no one else and was presumably not in general use. Another pigment mentioned only by Field is bone brown, which is made by charring bones in the same way as for ivory or bone black but for a shorter period.

Black Lead and Black Chalk

Many blacks are organic pigments made by burning or roasting a substance so that either the soot can be collected or the material itself can be used when it has turned black. There are, however, a few black or dark grey substances which can be used as pigments without such preliminary treatment, amongst which are black lead, black chalk and coal, all of which contain carbon.

The first should probably be known as graphite even though it is frequently called black lead. It is listed in very few documentary sources on painting, being mentioned in *Limming,* 1573, by Haydocke and Bate, and in *Arts Companion,* 1749. When ground and tempered with gum water, the colour was fairly transparent and suitable for use as a water colour as long as a strong black was not required.

Black chalk, which was probably used even less than graphite, is generally regarded as a clay containing carbon. It is listed in MS. Ashmole 1494, p. 636, and it is mentioned by de Mayerne in B.M. MS. Sloane 2052, f. 4, as a black which dries well, is easily extended and is a more valuable pigment in oil than common coal. His comment concerning drying suggests that the pigment which he called black chalk contained a proportion of something other than slow-drying carbon, possibly manganese or iron. However, his view concerning its value is most exceptional, and coal is mentioned much more frequently than black chalk in seventeenth-century sources.

Coal

The rock which is commonly called coal contains carbon, hydrogen and oxygen and is formed organically from vegetable matter. It is mentioned as a black pigment in a number of English sources dating from the beginning of the seventeenth century, largely perhaps because coal had by that time acquired importance as fuel owing to a shortage of timber. The term *sea coal* was frequently used, whether in reference to open-cast mining on the coast or to its transport route is not clear, and it is specifically mentioned in several books on painting. The name appears in MSS. Ashmole 1494, Sloane 6284, (where *smyth coale* is also mentioned) and in Stowe 680 (where it is distinguished from *sallowe cole blacke*). It is also mentioned by Bate and Gyles. De Mayerne refers to it, in French, as common coal or Scottish coal. Obviously various qualifications helped to distinguish common coal from charcoal. The references listed indicate that coal was used to some extent as a black pigment in oils and water colours during the first part of the seventeenth century but hardly at all after that time.

Charcoal (Blue Black)

At one time there were two names for the product obtained by partially burning twigs or small branches. One, *charcoal,* is still in use today because it is still made by the same slow-burning process, but *smallcoal* is obsolete. A description of its preparation by a quick-burning process is included in Chambers' *Cyclopaedia* of 1728.[293] Smallcoal is mentioned as a pigment by Gyles and Marshall Smith, but is cited less frequently than charcoal, which is listed as an oil-colour black in V. & A. MS. 86.EE.69 and by Bate, and as a water colour by Haydocke and Hilliard. The last named mentions willow coal in addition to common charcoal, whereas several writers of the seventeenth and eighteenth centuries suggest that vine charcoal is best for a blue black. The blue cast of charcoal black was doubtless observed at an early period; it was the reason why Bate recommended it for painting shadows in white ruffs in oils and Gyles names it for painting the shadows in a face. The name, blue black, appears to have been generally accepted and used during and after the second half of the seventeenth century. Gyles gives the following instructions for a painter to make blue black himself:[294]

> To make a faire blewe blacke.
> Take the cuttings of a vine, and burne them in a Crusiple as you doe Ivory blacke and it will be a fine blew blacke for a face blew then small cole blacke. Note that you cannott conveniently burne vine cuttings out of a Crusiple but they will consume to ashes.

Marshall Smith adds the information that the pigment must be ground in water and used with a drying oil if it is not mixed with another colour. References to blue black suggest that the colour was used mainly in oil, for, although it can be used as a water colour, other blacks are mentioned more frequently for water-colour painting.

Fruit Stone Black

A black pigment, prepared by charring the stones of cherries, peaches or dates in a crucible in the same way as Gyles suggests for blue black, is often recommended for use as a water colour in early seventeenth-century sources. Burnt peach stones are specified in V. & A. MS. 86.EE.69, whereas Hilliard recommends both cherry and date stones. Norgate mentions cherry stone black as being good for draperies and apparel. Variations are supplied by Haydocke and Peacham, the former suggesting burnt almond shells and the latter burnt walnut shells. Such blacks were used mainly during the seventeenth century by those water-colour painters who prepared their own colours. They may have been used to a limited extent in oil at the same time, as de Mayerne mentions both peach stone black and ivory black as being extremely slow-drying in oil.[295]

Ivory Black: $C, Ca_3(PO_4)_2$

By most accounts the best and deepest black was made by charring ivory, horn or bone in a crucible. During the seventeenth century, ivory black was likely to have been made from ivory rather than bones, because waste cuttings or raspings were available from comb-makers. Alternatively, the painter need not always have obtained raspings from such a source but could have converted something of his own to ivory black as

Peacham suggests: 'For a shift you may burne an old combe, fanne handle, or knife haft, or any thing els that is iuory, they wil make a very good black in water, but in oyle the best of all others.'[296] Hilliard gives detailed instructions for burning ivory with a little salt in a crucible for a quarter of an hour. After cooling, the pigment must be ground with gum solution and some water poured on so that the gum scum which rises to the surface can be poured off. The remaining pigment should be dried and stored, and a small quantity tempered with gum water when the time comes to use it. De Mayerne also mentions salt in connection with the manufacture of ivory black although not in the detailed instructions attributed to Mytens.[297] However, almost all instructions, of which there are many, emphasise the fact that the pot or crucible must be closed or sealed so that it is virtually air-tight, thus ensuring that the contents are only partially burnt and not reduced to white ashes.

Ivory black is frequently mentioned as a good black in oil, its transparency making it particularly suitable for glazing. A disadvantage in oil is its exceptionally long drying time, which was frequently modified in the seventeenth century by first grinding the pigment in linseed or nut oil and then adding a small quantity of verdigris to it as a drier. Bardwell praises ivory black, saying that it sympathises and mixes well with all other colours and is a good shading colour for blue. A few later writers are a little more critical; for example, Williams states that it is somewhat greasy. It was a cheap colour, for it was manufactured commercially during the eighteenth century, probably from bones instead of ivory (mutton bones are mentioned in some sources), and Dossie observes that the colour is often adulterated with charcoal.

Ivory black was popular with miniature painters, although it was insufficiently transparent for some painters engaged in washing prints. The water colour required a certain amount of skill in preparation, for seventeenth-century writers issue warnings that too much gum will make it glossy, that sugar candy is required to prevent cracking, and one, Bate, includes the addition of neat's gall, suggesting by implication that the colour did not flow well.

A little trouble in the preparation of ivory black was obviously worth while for seventeenth-century limners, but later water-colour painters found Indian ink, which is a variety of lamp black, more suitable for broad washes.

Lamp Black

Soot is probably the easiest pigment for a painter to make for himself. Peacham explains how it is done.[298]

> Ordinary lamp black. Take a torch or a link, and hold it vnder the bottom of a latten basen, and as it groweth to bee furd and black within, strike it with a feather into some shell or other, and grind it with gumme water.

Lamp black was also made commercially; John Smith mentions that soot made by burning rosin is imported from Scandinavia, and Waller refers

Fig. 45. Eighteenth-century apparatus for the commercial manufacture of lamp black. The fire was arranged so that, as smoke passed up to the vent in the roof, soot was deposited in the conical canvas. A rope and pulley was operated to lower the canvas for collection of the carbon black particles (Diderot and d'Alembert, *Encyclopédie*, 1754–1765). (Reproduced by courtesy of the University of London Library.)

to its manufacture in rooms built so that the soot cannot escape. Such a building is illustrated in the encyclopaedia of Diderot and Alembert in the section on mineralogy (*Figure 45*). The illustration shows a circular, windowless building with a conical roof. Smoke was channelled into the room and drawn up towards a hole in the roof, passing through a canvas which collected the particles of soot. A pulley allowed the canvas to be lowered or shaken at intervals so that the soot could be collected.[299]

Several writers suggest that, before lamp black is mixed with oil, the pigment should be burnt to eliminate some of its greasy quality. They invariably state that a drier should be added to it to accelerate its exceptionally slow drying time. Whereas verdigris is often suggested as a drier for ivory black, umber is recommended for lamp black, probably because that pigment tends more to brown than ivory black. Lamp black water colour is mentioned less than oil colour, as, during the eighteenth century, it was used in the form of stick ink, generally described as Indian ink.

Indian Ink

Stick ink imported from China must have become an increasingly familiar material throughout the seventeenth century as direct trade developed between Europe and the East. An early reference to it occurs in one of de Mayerne's manuscripts which includes instructions for making stick ink:[300]

> To make Chyna incke
> Take lamblacke, and grinde it well on a painters stone with [water] wherein gummj Arabike hath beene dissolued and when it is well ground, putt it into your moulds, then let it stand till it be drye, then vse it.

Black ink made in such a way would not have been exactly the same as Indian ink, which contains some kind of fish content. Dossie states that it is made from burnt fish bones mixed with isinglass and gives a recipe for making similar ink using ivory black, isinglass and Spanish liquorice. Field repeats a nineteenth-century idea that it contains sepia. Most painters were not particularly concerned with its composition but praised Indian ink for its transparency and easy flow. Apart from de Mayerne's reference, it is mentioned by Goeree in the seventeenth century and also by a large number of writers of the eighteenth and nineteenth centuries. Sticks of Indian ink varied in size but they were fairly expensive, ranging from sixpence to five shillings each at the end of the eighteenth century. Modern examples are illustrated in *Figure 11*.

Printers' Black

A number of blacks can be produced by partially burning a variety of materials. Amongst those which have not already been mentioned are two pigments, lake and verditer, both of which are mentioned as a source of black in seventeenth-century books, and cork, which is mentioned by Field. Burnt cork produced a lighter and softer black than wine lees, which used to be burnt and used in the production of printers' ink. The printers' pigment was known as Frankfort [*sic*] or German black, since it was imported from there. According to a description in Chambers' *Cyclopaedia,* 1728, it was made from burnt wine lees to which some ivory or fruit stone black was added. It was also available in France, but the German product was preferred in England because there was a difference in the wine lees. Printers' black was easily obtainable during the seventeenth century and painters used it for washing prints on account of

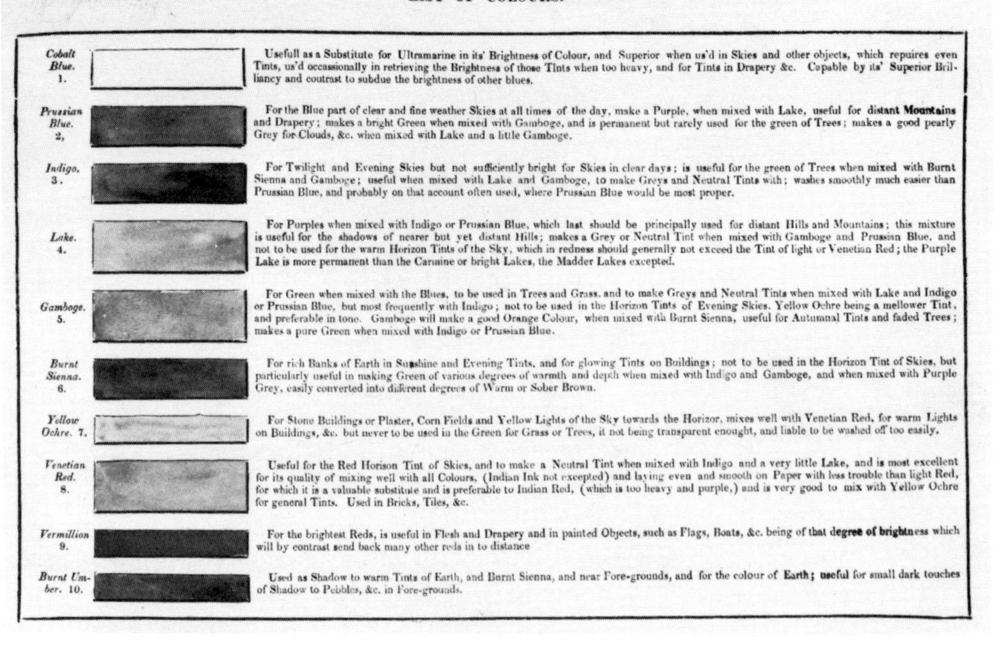

J. VARLEY'S,

LIST OF COLOURS.

Cobalt Blue. 1.	Usefull as a Substitute for Ultramarine in its' Brightness of Colour, and Superior when us'd in Skies and other objects, which repuires even Tints, us'd occassionally in retrieving the Brightness of those Tints when too heavy, and for Tints in Drapery &c. Capable by its' Superior Brilliancy and contrast to subdue the brightness of other blues.
Prussian Blue. 2,	For the Blue part of clear and fine weather Skies at all times of the day, make a Purple, when mixed with Lake, useful for distant **Mountains** and Drapery; makes a bright Green when mixed with Gamboge, and is permanent but rarely used for the green of Trees; makes a good pearly Grey for Clouds, &c. when mixed with Lake and a little Gamboge.
Indigo. 3.	For Twilight and Evening Skies but not sufficiently bright for Skies in clear days; is useful for the green of Trees when mixed with Burnt Sienna and Gamboge; useful when mixed with Lake and Gamboge, to make Greys and Neutral Tints with; washes smoothly much easier than Prussian Blue, and probably on that account often used, where Prussian Blue would be most proper.
Lake. 4.	For Purples when mixed with Indigo or Prussian Blue, which last should be principally used for distant Hills and Mountains; this mixture is useful for the shadows of nearer but yet distant Hills; makes a Grey or Neutral Tint when mixed with Gamboge and Prussian Blue, and not to be used for the warm Horizon Tints of the Sky, which in redness should generally not exceed the Tint of light or Venetian Red; the Purple Lake is more permanent than the Carmine or bright Lakes, the Madder Lakes excepted.
Gamboge. 5.	For Green when mixed with the Blues, to be used in Trees and Grass, and to make Greys and Neutral Tints when mixed with Lake and Indigo or Prussian Blue, but most frequently with Indigo; not to be used in the Horizon Tints of Evening Skies. Yellow Ochre being a mellower Tint, and preferable in tone. Gamboge will make a good Orange Colour, when mixed with Burnt Sienna, useful for Autumnal Tints and faded Trees; makes a pure Green when mixed with Indigo or Prussian Blue.
Burnt Sienna. 6.	For rich Banks of Earth in Sunshine and Evening Tints, and for glowing Tints on Buildings; not to be used in the Horizon Tint of Skies, but particularly useful in making Green of various degrees of warmth and depth when mixed with Indigo and Gamboge, and when mixed with Purple Grey, easily converted into different degrees of Warm or Sober Brown.
Yellow Ochre. 7.	For Stone Buildings or Plaster, Corn Fields and Yellow Lights of the Sky towards the Horizon, mixes well with Venetian Red, for warm Lights on Buildings, &c. but never to be used in the Green for Grass or Trees, it not being transparent enought, and liable to be washed off too easily.
Venetian Red. 8.	Useful for the Red Horison Tint of Skies, and to make a Neutral Tint when mixed with Indigo and a very little Lake, and is most excellent for its quality of mixing well with all Colours, (Indian Ink not excepted) and laying even and smooth on Paper with less trouble than light Red, for which it is a valuable substitute and is preferable to Indian Red, (which is too heavy and purple,) and is very good to mix with Yellow Ochre for general Tints. Used in Bricks, Tiles, &c.
Vermillion 9.	For the brightest Reds, is useful in Flesh and Drapery and in painted Objects, such as Flags, Boats, &c. being of that **degree of brightness** which will by contrast send back many other reds in to distance
Burnt Umber. 10.	Used as Shadow to warm Tints of Earth, and Burnt Sienna, and near Fore-grounds, and for the colour of **Earth; useful** for small dark touches of Shadow to Pebbles, &c. in Fore-grounds.

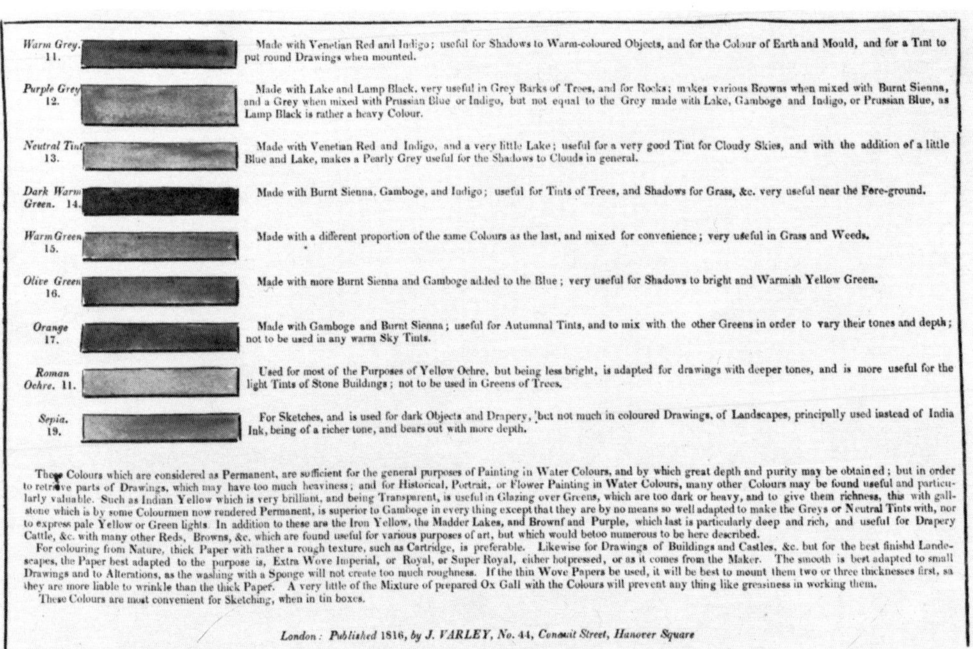

Warm Grey. 11.	Made with Venetian Red and Indigo; useful for Shadows to Warm-coloured Objects, and for the Colour of Earth and Mould, and for a Tint to put round Drawings when mounted.
Purple Grey 12.	Made with Lake and Lamp Black, very useful in Grey Barks of Trees, and for Rocks; makes various Browns when mixed with Burnt Sienna, and a Grey when mixed with Prussian Blue or Indigo, but not equal to the Grey made with Lake, Gamboge and Indigo, or Prussian Blue, as Lamp Black is rather a heavy Colour.
Neutral Tint 13.	Made with Venetian Red and Indigo, and a very little Lake; useful for a very good Tint for Cloudy Skies, and with the addition of a little Blue and Lake, makes a Pearly Grey useful for the Shadows to Clouds in general.
Dark Warm Green. 14.	Made with Burnt Sienna, Gamboge, and Indigo; useful for Tints of Trees, and Shadows for Grass, &c. very useful near the Fore-ground.
Warm Green 15.	Made with a different proportion of the same Colours as the last, and mixed for convenience; very useful in Grass and Weeds.
Olive Green 16.	Made with more Burnt Sienna and Gamboge added to the Blue; very useful for Shadows to bright and Warmish Yellow Green.
Orange 17.	Made with Gamboge and Burnt Sienna; useful for Autumnal Tints, and to mix with the other Greens in order to vary their tones and depth; not to be used in any warm Sky Tints.
Roman Ochre. 11.	Used for most of the Purposes of Yellow Ochre, but being less bright, is adapted for drawings with deeper tones, and is more useful for the light Tints of Stone Buildings; not to be used in Greens of Trees.
Sepia. 19.	For Sketches, and is used for dark Objects and Drapery, but not much in coloured Drawings, of Landscapes, principally used instead of India Ink, being of a richer tone, and bears out with more depth.

These Colours which are considered as Permanent, are sufficient for the general purposes of Painting in Water Colours, and by which great depth and purity may be obtained; but in order to retrieve parts of Drawings, which may have too much heaviness; and for Historical, Portrait, or Flower Painting in Water Colours, many other Colours may be found useful and particularly valuable. Such as Indian Yellow which is very brilliant, and being Transparent, is useful in Glazing over Greens, which are too dark or heavy, and to give them richness, this with gall-stone which is by some Colourmen now rendered Permanent, is superior to Gamboge in every thing except that they are by no means so well adapted to make the Greys or Neutral Tints with, nor to express pale Yellow or Green lights. In addition to these are the Iron Yellow, the Madder Lakes, and Brownf and Purple, which last is particularly deep and rich, and useful for Drapery Cattle, &c. with many other Reds, Browns, &c. which are found useful for various purposes of art, but which would betoo numerous to be here described.

For colouring from Nature, thick Paper with rather a rough texture, such as Cartridge, is preferable. Likewise for Drawings of Buildings and Castles, &c. but for the best finishd Landescapes, the Paper best adapted to the purpose is, Extra Wove Imperial, or Royal, or Super Royal, either hotpressed, or as it comes from the Maker. The smooth is best adapted to small Drawings and to Alterations, as the washing with a Sponge will not create too much roughness. If the thin Wove Papers be used, it will be best to mount them two or three thicknesses first, sa they are more liable to wrinkle than the thick Paper. A very little of the Mixture of prepared Ox Gall with the Colours will prevent any thing like greasiness in working them.

These Colours are most convenient for Sketching, when in tin boxes.

London: Published 1816, *by J. VARLEY, No.* 44, *Conduit Street, Hanover Square*

Fig. 46. *J. Varley's List of Colours*, 1816, in which red, brown and purple madder are cited but not illustrated. The area for cobalt blue is blank in this copy. (Reproduced by courtesy of the Victoria & Albert Museum.)

its good degree of transparency. References to it decrease as those to Indian ink become more numerous, suggesting, therefore, that Indian ink superseded printers' black for artists' use.

Greys

Hardly any grey pigments are listed in documentary sources, largely because grey can be made so easily by mixing other colours, and some pigments which may be regarded as grey are generally described under a different hue. For example, the least intense blacks, charcoal and coal, may be thought of as grey, and ultramarine ashes may be described as grey or very pale blue. The most obvious way of making grey is to mix black and white, but in the early nineteenth century a number of other mixtures were suggested for making a more subtle grey or neutral tint. Amongst such mixtures are Smith's grey, which is mentioned by Alston as a composition of lake, yellow ochre and Prussian blue, and Payne's grey, described by Alston as a mixture of lake, raw sienna and indigo. Varley's warm grey, purple grey and neutral tint are mentioned in his *List of Colours*. Mixed greys such as these were considered particularly suitable as water colours.

12

Whites

Bone White [$Ca_3(PO_4)_2$] and Shell White [$CaCO_3$]

Various pigments, composed mainly of calcium, were used occasionally in place of lead white, for they fulfilled a need for an inert white that was compatible with orpiment and did not darken on exposure when used in water-colour painting. Lead white, which was undoubtedly the most important and widely used white pigment, was deficient in both respects.

Bone white, which is composed of calcium phosphate and calcium carbonate, is mentioned occasionally in seventeenth-century sources although it is obvious that it was never used extensively. Bone white is listed in B.M. MS. Stowe 680, f. 133v, and burnt lambs' bones are mentioned by Hilliard. Writing in the eighteenth century, Dossie recommends calcined hartshorn or bone white as the only white useful for good water-colour painting.

Similarly, white made from eggshells or seashells is mentioned infrequently. Eggshell white is mentioned in B.M. MS. Sloane 6284, f. 109, by Dossie and by de Massoul, who explains that the shells must be crushed, boiled in water with a little quicklime, washed in clean water, crushed more finely, ground under a muller and finally allowed to dry in the sun.

An alternative white could be made by crushing either pearls or the best parts of oyster shells. The white obtained by this method is recommended in *The Art of Drawing,* 1731, and is mentioned by Dossie. Field also refers to it, saying that it is very white and has quite good body when used as a water colour but that true pearl white must not be confused with the cosmetic sold under the same name, which was made of bismuth.

Chalk: $CaCO_3$

Amongst naturally occurring sources of white pigments based on calcium are gypsum (hydrated calcium sulphate) and chalk (calcium carbonate). The former is rarely mentioned in documentary sources, appearing only in *Limming,* 1573, and Haydocke's translation of Lomazzo, where the point is made that, unlike lead white, it is compatible with orpiment. The same author mentions powdered marble, a crystalline form of calcium carbonate, as a pigment, but most writers refer to other varieties of calcium carbonate as chalk, whiting or white earth.

Some writers refer to white earth by its place of origin, and a number of such names were collected by Field. His list is given here with an indication of other writers who used the same colour names: Spanish white (MS. Stowe 680, Peacham, Norgate, Dossie, Watin, de Massoul), Troy white (Dossie), Rouen white (de Massoul), Bougeval white (Watin), Paris white, and China white (*The Art of Drawing,* third edition, 1732). An earth from Germany, *terra Goltbergensis,* is mentioned in the *Practical Treatise,* 1795, but not by Field. According to many accounts, most of the names listed genuinely indicated a finely powdered, soft white earth brought from the place mentioned. Most were obtained quite easily and cheaply, except for the Chinese variety which was scarce and little known.

Exceptionally, however, Spanish white was not necessarily Spanish, for several seventeenth-century writers give instructions for making it. Peacham and Norgate agree that it is made up of two parts of chalk to one of alum. The instructions given by Peacham are as quoted below.[301]

> There is an other white called Spanish white, which you may make your selfe in this manner; take fine chalk and grind it with the third part of Alome in faire water, till it be thick like pap, then roule it vp into balls, letting it lie til it be dry, when it is drie, put it into the fire, and let it remayne till it bee red whot like a burning coale, and then take it out and let it coole: it is the best white of al others to lace or garnish beeing ground with a weake Gumme water.

Norgate points out that it is a good idea to mix Spanish white with lead white 'for it will binde it well together, and is good to heighten upon'.[302] Most references to Spanish white and other forms of chalk are made by writers on water-colour painting, although Gyles lists whiting as one of two oil colour whites, the other being lead white. Nevertheless, it is difficult to establish to what extent chalk was used in oils and water colours, because it was sometimes included in the pigment called ceruse.

Since the late seventeenth century the name *ceruse* has been used as a synonym for lead white, but at an earlier period the name was used for a colour which might contain a proportion of lead white but did not necessarily do so. A large number of documentary sources of the sixteenth and seventeenth centuries list both ceruse and lead white without indicating the difference betweeen them. It is clear that, during the sixteenth century, English writers would use the name *ceruse* indiscriminately for a white pigment made from lead or tin, whereas some seventeenth-century writers applied the name to a pigment in which lead white and chalk were combined. Definite indications that ceruse contained chalk are supplied by de Mayerne ('en la Ceruse commune il y a la moitié de croye'), by Browne, who states that *serus* contains two parts of chalk to one of alum, thus equating it with Norgate's Spanish white, and by Boutet ('le Blanc de Ceruse, où il y a de la Craye ou Blanc d'Espagne').[303] It is, therefore, reasonable to suppose that a mixture is meant by other writers of the early seventeenth century who mention both ceruse and lead white without explaining the difference. This is likely to be the case with Norgate, who remarks: 'Ceruse will many tymes after it is wrought tarnish, starve and dye, and that which yow lay on with your pencell for a ffayre white will within a few Monethes become Rustie, reddishe or Inclininge to a yellow.'[304] These comments are corroborated by de Mayerne, who states that ceruse tends to yellow.[305] Even so, it is impossible to state that late sixteenth-century and early seventeenth-century ceruse always contained chalk, for at an earlier period the name could indicate tin white and a little later it was used for pure lead white. The period during which it was applied to a lead white and chalk mixture was relatively short. For that reason, it is impossible to describe with certainty the composition of ceruse out of context, and, even when the pigment is considered in relation to other whites listed in the same source, it is frequently difficult to come to a definite conclusion about it.

There must be some doubt also as to whether or not some painters knew exactly what they were using when a pigment was sold under a fashionable but ambiguous name such as Venice ceruse. The name appears in many English manuscripts dating from the late sixteenth and early seventeenth centuries, and it is quite impossible to tell if it was in fact Venetian, a pure white lead or a mixture. Certainly, by the early eighteenth century the name meant nothing to one writer who, confronted by Boutet's description of *le blanc de ceruse de Venise* as a pigment containing chalk, merely translated it as flake white and omitted the explanation.[306] A modern view on the subject is that from the end of the seventeenth century Venetian ceruse was a mixture of barytes and white lead. Further reference to the naming of white lead and barytes mixtures is made below in the discussion concerning barytes.

Lead White: $2PbCO_3 \cdot Pb(OH)_2$

Lead white is a pigment with a long history, and it is, therefore, not surprising to find instructions for its manufacture in a large number of sixteenth-century and early seventeenth-century manuscripts, the earliest of which, B.M. MS. Sloane 122, includes a detailed explanation of the way in which a clean barrel should be prepared with two gallons of vinegar at the bottom and strips of lead suspended inside above the vinegar. After the barrel has been made as air-tight as possible, it should be left for eight weeks, when it can be opened up and the white pigment knocked off the lead strips.[307] In discussing methods for the manufacture of lead white in medieval times, Thompson points out that the stipulation that the container should be tightly sealed would mean that the white formation on the lead would be lead acetate and not basic lead carbonate, which is the generally accepted composition of twentieth-century lead white. Medieval sources suggest that lead white made by the method outlined above was generally roasted in the sun, by which means basic lead carbonate might have been formed.[308] Manuscripts from the early modern period which include instructions for making white in a barrel do not include instructions for roasting the pigment later, and they are also deficient in that they do not always include the information that the barrel should be set in or near a source of heat. Norgate's instructions for making lead white on a small scale are somewhat unreliable because he suggests that pieces of lead should be placed in a gallipot, immersed in vinegar and the whole covered with a lead lid.[309] This contradicts many other instructions which state quite clearly that the lead must be suspended above the vinegar, although Norgate's method would presumably have had some success, as fumes from the vinegar would have allowed white to form beneath the lead lid. However, unlike some writers of the previous century, he includes the suggestion that lead white pigment should be exposed to the sun for two or three days. The practice is also mentioned in both of de Mayerne's manuscripts which are devoted to art subjects, and roasting lead white over the fire to obtain a better white is mentioned in another manuscript which dates from the early seventeenth century.[310] Some form of roasting seems to have been a usual measure, and it is likely, therefore, that it was practised even though it is not mentioned in some sixteenth-century manuscripts. Omissions and inaccuracies in various instructions suggest that the writers had not necessarily carried out the instructions, which would not be surprising in view of the fact that a commercial product was available.

Lead white was manufactured on a fairly large scale in England in the seventeenth century, as, when a monopoly was granted in 1622 for making white and red lead, there were at least four places where the industry was already practised.[311] None of the three early patents in which lead white is mentioned gives any details of manufacture, but Vernatti provided the Royal Society with a written account of a factory in the second half of the seventeenth century. Lead was cast in plates measuring one yard long and six inches wide, each being sufficiently thin so that it could be rolled loosely. Pots were prepared with a quantity of vinegar at the bottom and a bar across so that the coil of lead which was placed inside would be held above the vinegar. A plate of lead was placed over each pot to form a lid. A square bed of dung was prepared and 400 pots were set in it in rows and covered with boards which allowed another batch to be arranged. In all, four layers were stacked and covered over, and 1600 pots containing the same number of lead coils were left standing for three weeks, after which time the lead covers and coils were removed and beaten to dislodge the white flakes. There was considerable variation between the amount of white obtained from different pots, and some yielded none at all. The lead covers commonly bore thicker and better flakes than the coils, the best white being obtained from the outside of the covers. When the flakes had been separated from the remaining metal, they were ground with water between millstones, and the pigment was then moulded into small pieces and exposed to the sun to dry.[312] Later accounts written during the eighteenth and nineteenth centuries are very similar to this detailed description of what is now known as the stack process. It is interesting to find that there is no

Fig. 47. Setting the beds for lead white: a nineteenth-century illustration of the stack process (Dodd, *British Manufactures: Chemical*, 1844). (Reproduced by courtesy of the University of London Library.)

suggestion that the pots should be tightly closed, and a later writer, Watson, remarks that the top of each pot is covered in such a way that vapour can escape but nothing can fall in.[313] Erratic results, which were a cause of remark in the seventeenth century, were probably brought about by variations in the way each pot was closed with a lead plate: some were closed too tightly for success, whereas others allowed acetic acid vapour to escape and carbon dioxide to enter and complete the chemical reaction, which resulted in the formation of a mixture of basic lead carbonate and lead acetate.[314] Hardly any alterations were made to the process in the eighteenth century, although a patent, number 1581, taken out in 1787, specified the use of tanners' bark instead of layers of dung because it would produce a more certain degree of heat which was required for vaporising the acetic acid and forming carbon dioxide.

Apart from commenting upon the erratic results achieved in the industrial process of lead white manufacture, seventeenth-century Fellows of the Royal Society were very concerned about the physical condition of the workmen, who suffered from abdominal pains, contortions, shortness of breath and more acute symptoms of lead poisoning, giddiness, blindness and paralysis. The workmen who suffered most were those occupied in beating the lead in order to separate the white flakes and those engaged in grinding and drying the pigment. Vernatti described the symptoms, and at a later meeting of the Royal Society a doctor who attended workers at a lead white factory at Hatton Garden described the conditions there.[315] The problem was equally serious in the eighteenth century, and, although in 1783 the Royal Society of Arts offered a prize for a harmless method of preparing lead white, no useful suggestions were made. However, during the 1790s, Walker, Ward & Co. of Islington used equipment which prevented workmen inhaling lead white while it was separated from the metal. It consisted of several brass rollers one above the other in a framework twelve feet long, six feet wide and almost four feet deep. There were holes in the wooden base so that when the apparatus was partly filled with water and lead from the stacks was put through the rollers, the white pigment was separated and forced down through the holes, thus preventing it from escaping into the air.[316]

A less serious risk of lead poisoning was encountered when lead white was ground in a painting medium. Seventeenth-century painters do not seem to have objected to the pigment in the same way as they disliked orpiment with its offensive odour which constantly reminded them of its ill effects. Some apparently ignored the danger of lead white completely. For example, Gyles repeats Norgate's instructions for laying a ground of lead white preparatory to painting a portrait miniature on card, but then adds his own comment to the effect that if it does not lie smooth 'you may lick it all of with your tounge or wipe it of with a moist spunge'.[317] As long as a painter did not make a habit of consuming the colour, he would probably not suffer unduly, whereas the risk of cumulative lead poisoning was a much greater danger for the workmen engaged in the full-time preparation of artists' colours. A French device for protecting men engaged in colour grinding is described in the Royal Society of Arts *Transactions* for 1796 (*Figure 48*). The grinding slab was placed on a box table with a small gap around the edge so that fresh air which was piped into the table rose through the gap, was drawn up into apparatus in the shape of an open-based pyramid and was extracted through a pipe. An overhead charcoal furnace connected with the apparatus ensured that the air was drawn straight up, taking with it any dangerous fumes or loose pigment particles. The pyramid shape was made of glass so that the colourman could look down through it while working with his slab and muller.[318] In the early nineteenth century an English invention, Rawlinson's hand-operated single-roll grinding mill (*Figure 14*), was recommended, not only because it enabled greater quantities of colours to be prepared at one time than was possible with the traditional method of hand grinding but also on the ground that the colourman was less exposed to the dangers of poisonous colours.[319]

There are numerous references to lead white in literature on painting. White lead was the usual name until the late seventeenth century when the name flake white came into general use, being listed in the English translation of Goeree's book and in the 1675 edition of Browne's work. Different editions of John Smith's book reflect changes in nomenclature. In the edition of 1676 he mentions flake white 'which is by some accounted the best white of all others, but the reason of that I don't well understand, except it be, because it is scarce and dear'.[320] Marshall Smith lists the colour as white flakes, saying that it should first be ground in water before it is tempered with nut oil and that for dead-colouring it may be ground with linseed oil. Eighteenth-century writers point out that it is

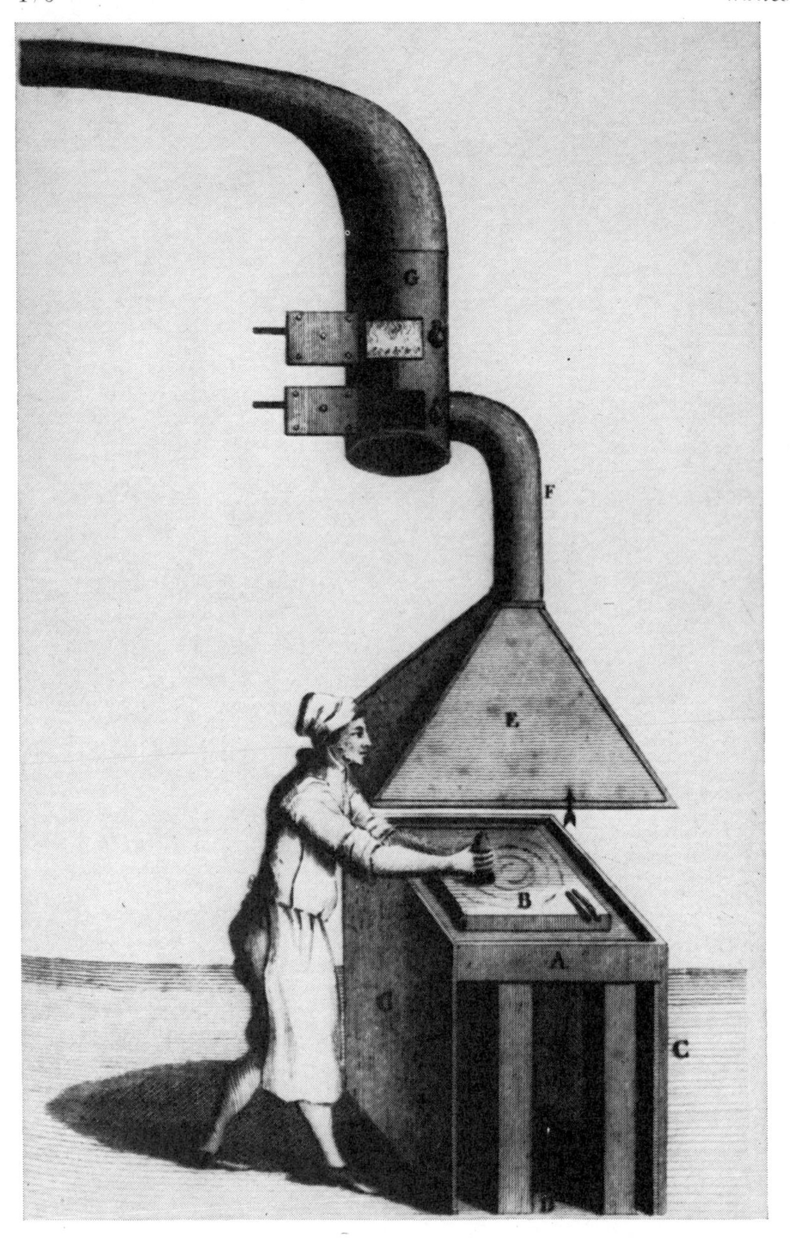

Fig. 48. Apparatus designed in France in 1796 for hand-grinding with the benefit of an air extractor to protect the colourman from toxic pigments: the grinding slab B stands on supports D which are surrounded by a box table A and C; a small gap between A and B allows fresh air drawn up by the charcoal fire at G to pass into the pipe F, so carrying away pigment particles and dangerous fumes. The workman was able to look down through the glass shield E (*Repertory of Patents of Invention*, v, 1796).

a good idea to purchase lead white in flake form because only then can one be sure that it has not been adulterated. Nottingham white is listed in addition to flake white in some books written towards the end of the century, including that by Williams, *The Artist's Repository* and *Practical Treatise,* the last of which explains that vinegar is used in the manufacture of usual lead white whereas alegar (sour ale or malt vinegar) is used for Nottingham white. Other names associated with basic lead carbonate were silver white and French white, which were synonymous according to Field. Ure states that silver white is very good because it is the finest pigment obtained from the final stage of the washing process. Field mentions Cremnitz white, otherwise known as Vienna white, as being imported in cube form. According to Ure, it used to be very good because the lead was particularly free from impurity, but the factory at Krems had closed down by 1839 when he was writing his dictionary of arts and manufactures. It is worth noting that the colour was misnamed in English, for it was called Cremnitz, which is the name of a town in Hungary, instead of Krems, the name of its place of origin in Austria.

Water-colour painters were not altogether deterred from using lead white despite the fact that when the pigment was bound in gum it would darken if exposed to air containing hydrogen sulphide. Norgate had a remedy for restoring discoloured white in the shell:[321]

> It is only Water, moste vsefull for all Collours (I meane in Lymning) but especially for recovering and preserving of white Leade or Cerouse tarnished, starved, deade or rusty with time or ill keeping. Take Rosemary water distilled, and with a fewe dropps of it temper your Shell of white soe starved and deade as aforesaid, and yow shall Instantly see it become perfectly white, and if for better proofe of this water, yow will only temper the one halfe of the Collour reserving the other vntoucht, yow will finde that parte yow haue moystned with this water restored to his pristinate puritie, the other remayning discoloured as before.

Painters continued to use lead-white water colour throughout the eighteenth century, regardless of warnings in many books that it would turn black in a month or two. At the end of the century Payne recommended that, if it was used in portrait miniatures, the painting should be glazed immediately, 'which method is the only one which stands a chance of preserving its purity'.

A few painters recommended the use of white with other colours to make opaque water colours. Ceruse, a mixture of lead white and chalk, was recommended by the French writer Boutet in the seventeenth century, and the recommendation was repeated in the English translation of the eighteenth century: 'Let it enter into all your Mixtures, in order to give 'em a certain Body, which will render your work glewish, and make it appear soft, plump, and strong.'[322] Such use of white was not universally approved, especially in England, and *The Art of Drawing,* 1731, practically forbids the use of white: 'While I am speaking of white for illuminating of Prints as I have already observ'd, that the clear white of the Paper is proper to be left uncolour'd.'[323] However, the subject was controversial, and a copy of the book in the Victoria & Albert Museum contains the following manuscript note, probably inserted during the second half of the eighteenth century:[324]

> This is not the most masterly manner. The best way and which gives the print the air of a painting is to load the light parts with a good body of colour, but this requires a hand of great judgement, and of abilities equal to a superior kind of painting.

It is clear that a number of painters in water colours used lead white throughout the period under discussion. At the end of the eighteenth and beginning of the nineteenth centuries, white precipitate of lead was used in water-colour painting. The *Practical Treatise,* 1795, contains full instructions for making the pigment, which Field describes as sulphate of lead, a white precipitate produced from any solution of lead by adding sulphuric acid. It was used only in water-colour painting, for it was considered by some more suitable than flake white for that technique. Field confirmed that, properly prepared, it was one of the best lead whites, but stated that some samples were inferior to flake white in body and permanence. It was still available in the 1830s, being sold in bottles under the name Flemish white.

Tin White

Until the early seventeenth century, tin white was used as a water colour, even though it was not brilliant and it had a reputation for turning grey. MS. Sloane 122, f. 92v, contains the statement that ceruse of tin is made in a barrel in precisely the same way as lead white, and instructions for making the latter reappear as instructions for making tin white in a manuscript of slightly later date. There are very few verbal differences between them, and all measurements are identical. The slightly later version which is quoted here is distinguished by the alchemical symbol for tin:[325]

> For to make Ceruse
>
> Take plates of tinne and beate them as thinne as thowe maist but lett ech plate be vij inches of length and ij of bredth, then take a Clene barrell that is sweete, and put therin ij gallons of viniger or of eysell and hange the plattes therouer in the barell so that none tuch the other, and stop fast the barelles mowth that no eyre com owte, and let it stand still viij weekes. After that open the barell softe, and take a stike and bete of the white that hangeth on the plates into a Clene vesell, after temper it up with veniger and put it in kekisse. Cakes.

B.M. MS. Sloane 1394, f. 137, contains basically similar instructions although verbal differences, especially differences in the dimensions of the tin plates (two inches by one instead of seven inches by two), suggest that the instructions were merely copied but not carried out by the seventeenth-century compiler.

Tin white, which, according to documentary sources, was used in manuscripts, was unsatisfactory, as addition of medium to the pigment led to colour alteration, probably as a result of a change in refraction. As manuscript illumination declined, tin white became obsolete, but it was still known in the early seventeenth century. De Mayerne describes how Van Dyck experimented with the colour, although his account is somewhat confused because at the start he refers to both tin and bismuth, whereas tin alone is mentioned throughout the rest of the passage. Van Dyck had tried tin white in oil and reported that the most usual white, that made from lead, was very much whiter than tin as long as it was well washed. Tin white had insufficient body and was useful only for manuscript illumination. The same conclusion was reached by Mytens, who had also tried tin white and found that it blackened in sunlight and spoiled white lead if the two were mixed. It was useless in oil and also in distemper if exposed to the air.[326] Further passages in another of de Mayerne's manuscripts refer to tin white for cosmetic purposes and as a vitreous colour.[327] It was in the latter form that tin white remained in use for artistic purposes, for, long after it had disappeared from the palette of miniaturists, the colour was frequently employed by enamellers. In the second half of the seventeenth century Boyle referred to the use of tin oxide for enamelling, saying that it was well known amongst chemists that when tin was calcined a white substance was produced.[328] Attention was once again drawn to tin white at the end of the eighteenth century after the French chemist, Guyton de Morveau, had conducted a series of experiments to find a good, inert, non-toxic white pigment which was suitable for oil and water-colour painting.[329] Tin white, which was prepared in various ways, was declared unsuitable as it tended towards yellow or blue, but, following his mention of the pigment, it was referred to by Watin, who admitted that he had not tried it, and by Field, who regarded it as a bad drier with poor body in oil.

Bismuth White

Bismuth was known in the early sixteenth century, possibly earlier, for it is mentioned in written sources on mining and metallurgy. Agricola, who called it *bisemutum* or *plumbum cinereum* (ash-coloured lead), described its metallic character and stated that it was used in alloys and for making a pigment.[330] However, bismuth was sometimes confused with other metals, such as tin and antimony, and reference has already been made to the way in which de Mayerne included bismuth in a passage concerning tin white. There the word *Wismut* is used, and the same word occurs subsequently in the same manuscript where the white precipitate of bismuth is described as an excellent white colour.[331] The pigment was easily prepared by obtaining a solution of bismuth by means of nitric acid and then adding water to the solution so that a white precipitate was formed. It seems unlikely that bismuth white was prepared commercially for artistic purposes, for it is hardly ever mentioned in literary sources, appearing under the name *flowers of bismuth* in the not very authoritative *The Young Artists' Complete Magazine, circa* 1785. Its inclusion there is probably because bismuth was prepared for cosmetic purposes. Various scientific writers pointed out that bismuth white was poisonous and quickly blackened on exposure to air, both features which should have excluded the pigment from painting and cosmetics, but it seems that eighteenth-century fashion required a pallid complexion, so there was a ready demand for a preparation of bismuth which was sold under the names *Spanish white* and *pearl white*.[332] For that reason it was also available for painting purposes for the few people who decided to use it.

Light White

Silver is mentioned occasionally in seventeenth-century sources, suggesting, therefore, that it was used to some extent in manuscript illumination of an earlier period. Shell silver is included in Goeree's list of whites, but it is obviously a metallic colour, that is, silver leaf ground finely, tempered with gum water and preserved in a shell. However, several eighteenth-century sources include instructions for making a white from silver for use in miniature painting. Instructions are provided in the 1757 edition of *The Art of Drawing* and by de Massoul who calls it light white.[333]

> To make this White, take a sheet of silver, which you beat as thin as paper; cut it in pieces about the size of a halfpenny; steep it in Aqua Fortis for twenty four hours; being dissolved, pour off the Aqua Fortis, and wash what is left at the bottom of the vessel, five or six times in distilled water, till there be no remains of Aqua Fortis left in the dissolution; which may easily be known by touching it with your tongue. It is afterwards dryed.

Ackermann sold the colour under the name *light white*, and another colour called Ackermann's white is also listed under silver in his book on water colours, although it seems possible that the latter was composed of zinc as well as silver. White made from silver or quicksilver is mentioned in Guyton de Morveau's paper on artists' whites. He explains that a white precipitate can be obtained from either by dissolving the metal in nitric acid and adding a vegetable alkali, but both tend to alter rapidly on exposure to air.

A noteworthy feature of silver white is the potential confusion of names associated with it. In English, confusion with lead white is possible but unlikely, as the name *silver white* appears to have been applied exclusively to lead white whereas *light white* was the name reserved for silver. Great care is required, however, in the translation of French names, for *blanc d'argent* was sometimes used for zinc white, as in *Le maître de miniature,* 1820, and sometimes for lead white, as Bouvier used it as a synonym for Cremnitz white. *Blanc léger* is probably the correct translation of the English *light white*. In *L'art de peindre les fleurs* by Augustine Dufour the colour called *blanc léger* is described as being sold for miniature painting, but its composition is not identified. Reference to the original French editon of de Massoul's book (unfortunately not available for consultation by the writer) would show if *blanc léger* was the French name for white made from silver. If so, it would suggest that a recent interpretation of *blanc léger* as zinc white is incorrect.[334]

Barytes: $BaSO_4$

Barytes and heavy spar are two common names for barium sulphate, a mineral which occurs in numerous places, including the British Isles. According to a modern authority, Agricola was the first writer to describe barytes in the sixteenth century.[335] Full documentation concerning the mineral did not take place until the second half of the eighteenth century when, in 1774, Scheele encountered it in the course of investigation on the nature of native manganese dioxide and, a little later, his work was repeated by Guyton de Morveau, thus enabling the latter to include it in his examination of various white pigments.

Reference to English sources suggests that barium sulphate came into use as an artists' colour towards the end of the eighteenth century. In the nineteenth century, Ure reported that it was sold under the following names: Venice white (equal parts white lead and barytes), Hamburg white (one part white lead to two parts barytes) and Dutch white (one part white lead to three parts barytes).[336]

Neither native barytes nor manufactured barium sulphate was used alone as an oil colour, for, as Guyton de Morveau pointed out, the pigment itself has the appearance of a good white but as soon as it is tempered with oil it becomes grey and translucent and does not regain its colour even when the paint is dry. The translucency of the pigment makes it very suitable for use as an extender with oil colours, for a considerable quantity may be added to a colour without making it appear very much paler, whereas a relatively small quantity of lead white reduces any other colour to a pale tint because of its opacity. Nevertheless, painters were not generally interested in obtaining an extender, and barium sulphate was rejected by oil painters because it lacked opacity. It was adopted in water-colour painting, however, and the following remarks are included by Parkes in his *Chemical Essays,* 1815, which contain a complete paper on the history and use of barium sulphate:[337]

> The artificial sulphate of barytes has for many years been applied to the purposes of painting in water colours, and is the most delicate and perfect white in use. It was first recommended and brought forward by Mr Hume of Long Acre, who has long supplied the public with it under the name of *permanent white.* The native and artificial sulphate of barytes are both poisonous.

The colour was also called constant white, a name which was used more frequently than permanent white and appeared as early as 1783 in the sixth edition of *Bowles's Art of Painting in Water-Colours.* The same name appears in Hamilton's colour list, although he still maintains that flake white is the best white, and it is listed by Field, who suggests that as long as it is properly prepared and free from acid it is one of the best whites for water-colour painting. He lists permanent white, barytic white and constant white as synonyms. The main advantages of the colour were durability when exposed to light and air, and compatibility with all other colours. Translucency presented a problem during painting because the paint was translucent and low in tone when wet, but it became more opaque and therefore brighter as it dried. Marked alteration in tone could spoil the whole composition, especially in a large water-colour painting, and it was probably for that reason that constant white was used mainly in miniatures, for there bright touches of white were acceptable for highlights.

The historic importance of barium sulphate as a water colour has to some extent been overshadowed by its modern role as an extender for oil colours and also by the fact that zinc oxide was finally accepted as the best water colour white in the nineteenth century and so became better

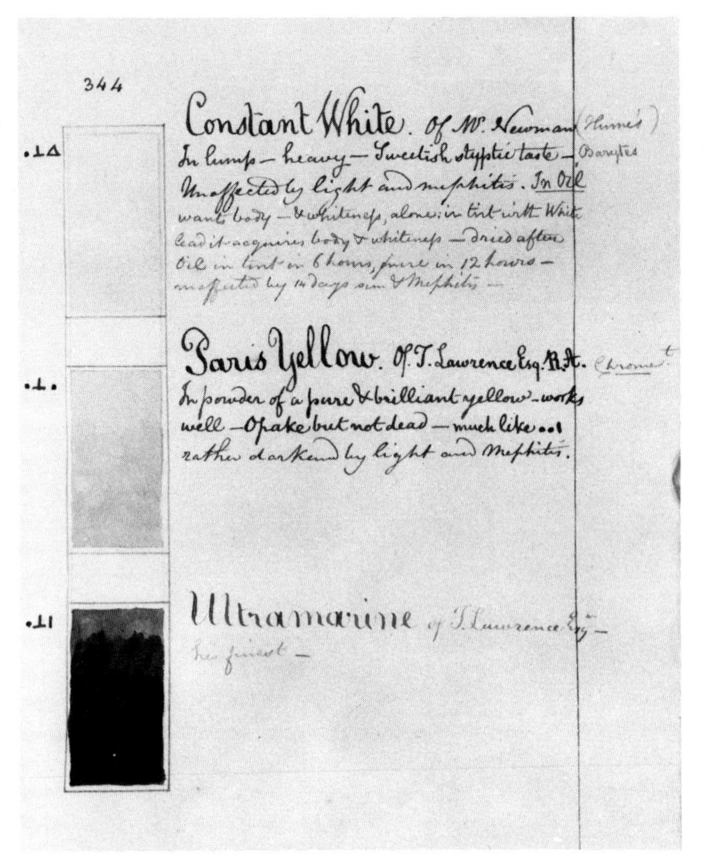
344

Constant White. of Mr. Newman (Hume's)
In lump — heavy — Sweetish styptic taste — Barytes
Unaffected by light and mephitis. In Oil
want body — & whiteness, alone; in tint with White
lead it acquires body & whiteness — dried after
Oil in tint in 6 hours, pure in 12 hours —
unaffected by 14 days sun & Mephitis —

Paris Yellow. of T. Lawrence Esq. R.A. Chromate
In powder of a pure & brilliant yellow — works
well — Opake but not dead — much like
rather darkened by light and Mephitis.

Ultramarine of T. Lawrence Esq. —
the finest —

Fig. 49. Field's comments on constant white (barium sulphate) obtained from the colourman Newman before 1815. Also shown is Paris yellow (lead chromate) and natural ultramarine obtained from the artist Thomas Lawrence. (Field, 'Practical Journal 1809', Courtauld Institute of Art MS. Field/6, f.344.) (Reproduced by courtesy of the Courtauld Institute of Art.)

known. At the turn of the century, however, water-colour painters were familiar with barium sulphate as constant or permanent white, and a number used the colour. Constant white was available for a considerable time after the establishment of zinc oxide, so that it served as a water colour of some importance for about a century.

Zinc White: ZnO

An opportunity to include zinc oxide in the artists' palette existed for some time before it was recommended for the purpose. Agricola was able to distinguish zinc ore *(lapis calaminaris)* and metallic zinc *(cadmia metallica),* and in his time the use of zinc in an alloy with copper to form brass was well known. In the seventeenth century two British writers, both Fellows of the Royal Society, described the occurrence and treatment of the ore. Edward Brown visited a mine in Limbourg, said to be 300 years old, where zinc ore was dug out, washed and then calcined in the open.[338] Merret mentioned the occurrence of *lapis calaminaris* in Somerset and North Wales and described how it was calcined in a reverberatory furnace for about five hours, during which time it was raked over. When properly calcined it was a fine, white powder: 'Almost half of the Calamie (as the workmen call it) is wasted and flies away in flour, which sticks to the mouth of the Furnace of divers colours of little use with them.'[339] Merret identified the powder as the basis of an ointment used in ancient times and passed on the information to Harvey, the famous physician, who used zinc ointment with success in medical practice. The white oxide was again mentioned by Shaw in 1733: 'The Inflammability of zink is very remarkable; for it burns durably of a bluish white Flame in the Fire; and thus resolves into a white Calx.'[340] During the first half of the eighteenth century several chemists conducted experiments on zinc after Henckel had isolated the element in 1721. Europeans, who had previously relied upon China and the East Indies as the main source of metallic zinc, began to extract the metal from zinc ore on a large scale, the production of zinc oxide being an incidental part of the process.

There was, therefore, plenty of opportunity to try zinc oxide as a pigment, but there is no mention of any experiment until 1782, when Guyton de Morveau reported on the scientific examination of all known white pigments and raw materials that might serve as a white pigment. He claimed that zinc oxide was a non-poisonous, inert pigment of good covering power and that it was the best alternative to the poisonous

Fig. 50. Pierre Adolphe Hall: self-portrait dated 1793. Zinc oxide has been identified in this miniature by V.J. Murrell as a result of its characteristic fluorescence under ultra-violet light. (On loan to the Victoria & Albert Museum from the National Museum, Stockholm.) (Reproduced by courtesy of the National Swedish Art Museums.)

pigment, lead white. It was a little more expensive than lead white, but he considered that painters should overlook that point as its advantages were so great, concluding his paper 'on ne marchande pas l'immortalité'. When the report was made, zinc oxide was already being manufactured commercially by Courtois, the laboratory steward of the Dijon Academy, who had found that the oil colour was slow drying but had remedied the matter by adding zinc sulphate as a drier. The pigment was sold at a price of six francs per pound.[341] Zinc white received a certain amount of attention, for the paper by Guyton de Morveau was reprinted in French and abstracts appeared in English, as, for example, in the *Practical Treatise,* 1795. The colour was mentioned in passing in books on painting, but none of the eighteenth-century writers had used it. Watin states that the use of zinc oxide and alum is recommended but it is too expensive for decorators. The following passage appears in an English book dating from the 1780s:[342]

> Zink has lately furnished an elegant white, which, if all said of it by the French, who prepare it, is true, is a noble production; as it stands perfectly, both in oil and water. It is not yet used in England.

However, manufacture in England was not delayed for long, as the specifications for two patents taken out by John Atkinson, a colour-maker of Harrington near Liverpool, are dated 1794 and 1796.[343] Zinc white was included amongst the colours available at the close of the century from the London factory of de Massoul, but still the pigment was not adopted. The claim that zinc oxide was non-poisonous and inert was perfectly true, but a chemist's pronouncement that the pigment had sufficient covering power was not supported by painters, with the result that many writers of the early nineteenth century comment on the inadequate body of zinc white in oil. Ibbetson's remark is just one example: 'White lead is the only white we have of sufficient body to use in oil. White has been made from zinc, but it has not sufficient substance.'[344] It was for that reason that zinc white was rejected by oil painters.

Colourmen were aware of the shortcoming of zinc oxide, for de Massoul mentions that it lacks body and, in consequence, it is often mixed with white prepared from silver for use as a water colour. In view of the fact that much of Ackermann's treatise of 1801 is based on the writing of de Massoul and he includes *white of zinc* and *chemical white* in the general list of pigments classed according to composition, it is surprising to find that neither name is included in the price-list of water-colour cakes. However, Ackermann's inclusion of two whites made from silver has already been mentioned, and, as both of them are included in the price-list and de Massoul suggests that zinc oxide is best mixed with precipitate of silver, it seems possible that the colour named Ackermann's white was a mixture of the two. Ackermann described it as 'a new colour, superior to any White hitherto known, both in washing it, and also in its durability'. But it cannot have met with much success because in *Ackermann's Manual* of 1844 permanent white (barium sulphate) is recommended as the best water-colour white. An entry dating from after 1815 in Field's 'Practical Journal 1809', f. 350, contains an example named New White, obtained by Field from Sir Thomas Lawrence, that seems very likely to be zinc white.

It was not until the 1830s that zinc white was finally established as an indispensable water colour. By tradition, the special form of zinc oxide which Winsor & Newton sold under the name Chinese white was introduced in 1834. The reason for the choice of name is unknown; there may perhaps have been an association of ideas between a pure white artists' colour and the excellence of the oriental porcelain known as Chinese white which was popular in Europe in the eighteenth century. The name was not immediately adopted by others, as it does not appear in Field's *Chromatography,* where zinc oxide is described as 'more celebrated than used'. In 1837 Bachhoffner dismissed zinc oxide in his book

14 WINSOR & NEWTON,

CHINESE WHITE.

A PREPARATION OF WHITE OXIDE OF ZINC,

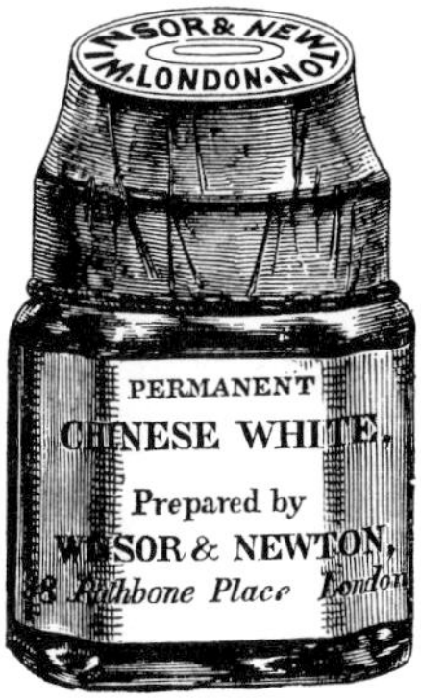

THE MOST ELIGIBLE WHITE PIGMENT FOR WATER COLOUR PAINTERS.

SOLD IN BOTTLES OR TUBES.

Professional price 1s. each. Retail Price 1s. 6d. each.

The White Oxide of Zinc is pronounced by the highest chemical authorities to be one of the most unchangeable substances in nature. Neither impure air, nor the most powerful re-agents, affect its whiteness. It is not injured by, nor does it injure, any known pigments.

It has long been pointed out by chemists as a most desirable substance for the Artists' use, provided sufficient body could be imparted to it; but until lately the want of this necessary quality rendered it unavailable. In WINSOR and NEWTON's preparation, termed Chinese White, this desideratum has been attained. The Chinese White, by combining body and permanency, is rendered far superior to those whites known as "Constant" or as "Permanent" Whites; and not having their clogging or pasty qualities, it works and washes with freedom.

The great body it possesses gives it the property of drying on paper of the same tone as it appears when first laid on, and thus, when used, either alone or in compound tints, it does not deceive the Artist like other whites, by drying up three or four tones higher than when wet.

The Chinese White is peculiarly available in mixing with any of the Water Colours in use, and particularly with the Moist Colours, thereby forming at pleasure an extensive range of body colours of a very superior kind.

The following Paragraphs are extracted from Mr. HARDING's *"Principles and Practice of Art."*

"When the Oxide of Zinc, which is prepared by Winsor and Newton under the "name of Chinese White, was first put into my hands, some years ago, I applied to "one of my friends, whose name as a chemist and philosopher is amongst the most "distinguished in our country, to analyze it for me, and to tell me if I might rely "on its durability; the reply was, that if it would in all other respects answer the "purposes I required of it, I had nothing to fear on account of its durability."

"This is an invaluable pigment." "It is hardly possible to overrate the value of "Opaque White in Water Colours when judiciously used."

Fig. 51. Catalogue entry for Chinese white in the 1840s. Winsor & Newton provided a new name and improved water colour formulation for a pigment that had been available for the previous fifty years. (Reproduced by the kind permission of Messrs Winsor and Newton Limited.)

Chemistry as applied to the Fine Arts saying that it was used occasionally but with little success, 'being much inferior both in body and colour to the whites with a base of lead, and but little, if any, superior to them in durability'. Bachhoffner was a chemist who gave popular lectures and also acted as a colourman, selling precipitate of lead under the name Flemish white. The manufacturers of Chinese white took exception to his condemnation of zinc oxide on the grounds that, whereas scientists would be sufficiently knowledgeable to ignore his remarks, painters without scientific knowledge would accept them as authoritative and would cease to use Chinese white. Their reply to Bachhoffner was printed, and copies were presumably circulated amongst water-colour painters who had formed the basis of their clientele throughout the first five years of their firm's existence.[345] The partners' defence of zinc white was clearly successful, as the manufacture and use of the colour continued and the trade name, Chinese white, was established as a synonym for zinc white water colour.

The lapse of fifty years between a chemist's recommendation of zinc oxide as a pigment and the successful and uninterrupted marketing of the artists' colour on a commercial scale is sufficient to prove that painters were not prepared to accept any new colour on the basis of theoretical advantages. To them, practical considerations were all important, and it is clear that, after intermittent trials over a considerable period, many painters preferred to retain the use of lead white, even in water colour, owing to its ideal working properties. To be told that zinc white was superior was not sufficient; it was not until effective measures were taken to improve the working properties of zinc white in water colour that it was finally accepted.

13

Science and Art

The legacy of the Middle Ages was still much in evidence in the sixteenth and seventeenth centuries; not only did painters retain traditional techniques but also many traditional pigments, some of which—ochres, vermilion, lead white and others—are still in use today. At that time, however, Europeans found distant parts of the world accessible and new technological processes attainable, in consequence of which painters were offered an opportunity to extend their colour range. When Bacon wrote 'arts and sciences should be like mines, where the noise of new works and further advances is heard on every side', changes seemed to take place far too slowly, but even then, at the very beginning of the seventeenth century, painters already benefited from developing technology.

New pigments were produced in industries which were otherwise unconnected with painting. In the sixteenth century the development of mining and glass-making techniques ensured that smalt was available throughout Europe. At a similar period the industrial use of mineral acids achieved importance. Nitric acid was used in silver refining, and the refiners produced blue and green verditer as a side-line. Hydrochloric acid was first available in the middle of the seventeenth century, and it was utilised a little later in the manufacture of Prussian blue. Sulphuric acid, much in demand for use in textile bleaching in the mid-eighteenth century, was important in the manufacture of ferrous sulphate and iron oxide pigments. Alchemists were undoubtedly capable of making iron oxides, and possibly other non-traditional pigments, but the commercial availability of all new colours was entirely dependent on economic circumstances; that is, they were manufactured if the process was part of a larger industry or if production could be maintained on economic lines because the pigment was required in activities other than painting. Thus, verditer was a by-product, smalt had application in the paper industry and laundry-work, and Prussian blue was used in textile dyeing. Those new colours, and manufactured iron oxides, were used in decorating as well as artistic painting.

Even at the end of the eighteenth century, when the pace of progress gathered momentum and alchemy gave place to chemistry which was founded on theory and aided by a new system of nomenclature far superior to the international but often ambiguous alchemical symbols, the availability of new pigments still depended upon economic circumstances. The potentialities of compounds made from several of the newly discovered elements were often recognised but not always exploited immediately. Occasionally a pigment could not go into commercial production at once because of a lack of the raw material; often the introduction of a pigment remained dependent on demand in other industries. A survey of artists' colours available in 1835 and their subsequent fate underlines the economic problem.

A good selection of blue pigments was available at that time. Natural ultramarine, azurite, smalt, Prussian blue and indigo had been joined fairly recently by cobalt blue, initially introduced for the ceramic industry, and French ultramarine. Together the last two replaced smalt and natural ultramarine, and with Prussian blue, and to a lesser extent indigo, served as the most important artists' blues of the nineteenth century. All had applications in other industries, but only three of the pigments are now available, because, when the natural product was replaced by synthetic indigo for textile dyeing, and indigo cultivation dwindled, supplies for artists ceased.

Green pigments were fewer in number than blues, and, although artists had no brilliant, inert green, their lack may not have worried them unduly as they could depend on mixed greens. Apart from terre verte and the organic colour, sap green, both of which grew in popularity in the eighteenth century, most green pigments were based on copper. None was ideal, and some were more unsatisfactory than others. The inventions of the late eighteenth century, Scheele's green and Brunswick green, were short-lived, so in 1835 malachite, verdigris and the new pigment, emerald green, were the main copper greens available. All three are now obsolete, although emerald green, which was used as an insecticide as well as a pigment, survived until the second half of the twentieth century. The most important research connected with green pigments in the early nineteenth century was undoubtedly the examination of chromium compounds. Opaque oxide of chromium found ready acceptance in the ceramic industry, and a transparent variety was made for artists a little later. Both are available as modern artists' colours, as also is the long-standing terre verte.

Yellow pigments offered a wide choice in the early nineteenth century. Massicot was obsolete, but orpiment and Naples yellow were still available although little used. The introduction of patent yellow, a by-product of the soda industry, had done much to replace orpiment, but it was itself being replaced by other pigments. The isolation of chromium, barium, platinum and cadmium made possible the manufacture of several new yellows. Initially, chrome ore was difficult to obtain, but once the raw material was located there was little delay in production because manufacture was stimulated by use of chrome yellow in calico printing. Platinum was used in ceramics, but, even so, the raw material remained too costly for artists' use, and cadmium too was hard to come by. Chrome yellow survived despite its deficiencies, because it could be made at an economic cost. Organic yellows available at the same time include Indian yellow, gamboge, which was imported in quantity for pharmaceutical as well as painting purposes, and quercitron, which was utilised in textile dyeing until replaced by synthetic organic yellows at a later period. Of the yellow pigments available in or soon after 1835, yellow ochre, Mars yellow, the chromes (including lemon yellow), cadmiums and gamboge are still supplied in the same form. Gold, also, has been retained as a water colour for illuminating in a bygone tradition.

A less extensive selection of red pigments existed. Red ochre, Mars red and vermilion were used, as were red lead and realgar to a lesser extent. Chrome red was introduced but never became as important as the yellow and orange varieties. Iodine scarlet also appeared as the result of scientific research, but the fact that iodine was used as an antiseptic soon after its discovery did not prolong the existence of an unsatisfactory pigment. A number of organic raw materials were used for making red, of which madder and cochineal were undoubtedly the most important, with safflower a poor third. The greatest achievement in connection with red pigments in the early nineteenth century was the improvement of lake-making technique, which may have been assisted by greater understanding of the use of metals in textile dyeing during the previous century. Improved technique enabled nineteenth-century colour-makers to produce an excellent range of madder colours from rose, through scarlet and crimson, to purple and brown. Documentary evidence suggests that the variety of madder lakes available then has been unparalleled before or since, for, with the subsequent introduction of synthetic alizarin and the consequent decline in the cultivation of madder, artists' colours made from the plant have become more costly and fewer in number. Rose madder is still available, however, as are the ochres, Mars colours, vermilion and carmine.

Black and brown pigments reflect the advance of science less than others, although they show the influence of fashion rather more—the use of asphaltum, mummy and bituminous earths demonstrating the desire of many artists for very translucent shadows in oil painting. Pigments which were available in 1835 and are still in use today include ochre, umber, Mars brown, Vandyke brown, ivory and lamp black, and Indian ink. Asphaltum, mummy, bistre and genuine sepia are now obsolete. White pigments are more interesting in that they are more closely associated with scientific discoveries of the eighteenth century. There were three important white pigments in 1835: lead white (the only one used in oil), zinc white and barytes. The last two were relatively new and had supplanted the traditional whites of previous centuries which were based on calcium. All three are still in use, although barytes is now restricted to acting as an extender, often to improve the working properties of another pigment.

Scientific papers are extremely important as sources for the history of the pigments which were introduced in the late eighteenth and early nineteenth centuries. Not only do they offer details concerning the way the pigments were made and also indicate other industries in which the colours were used, but they often provide information about the earliest

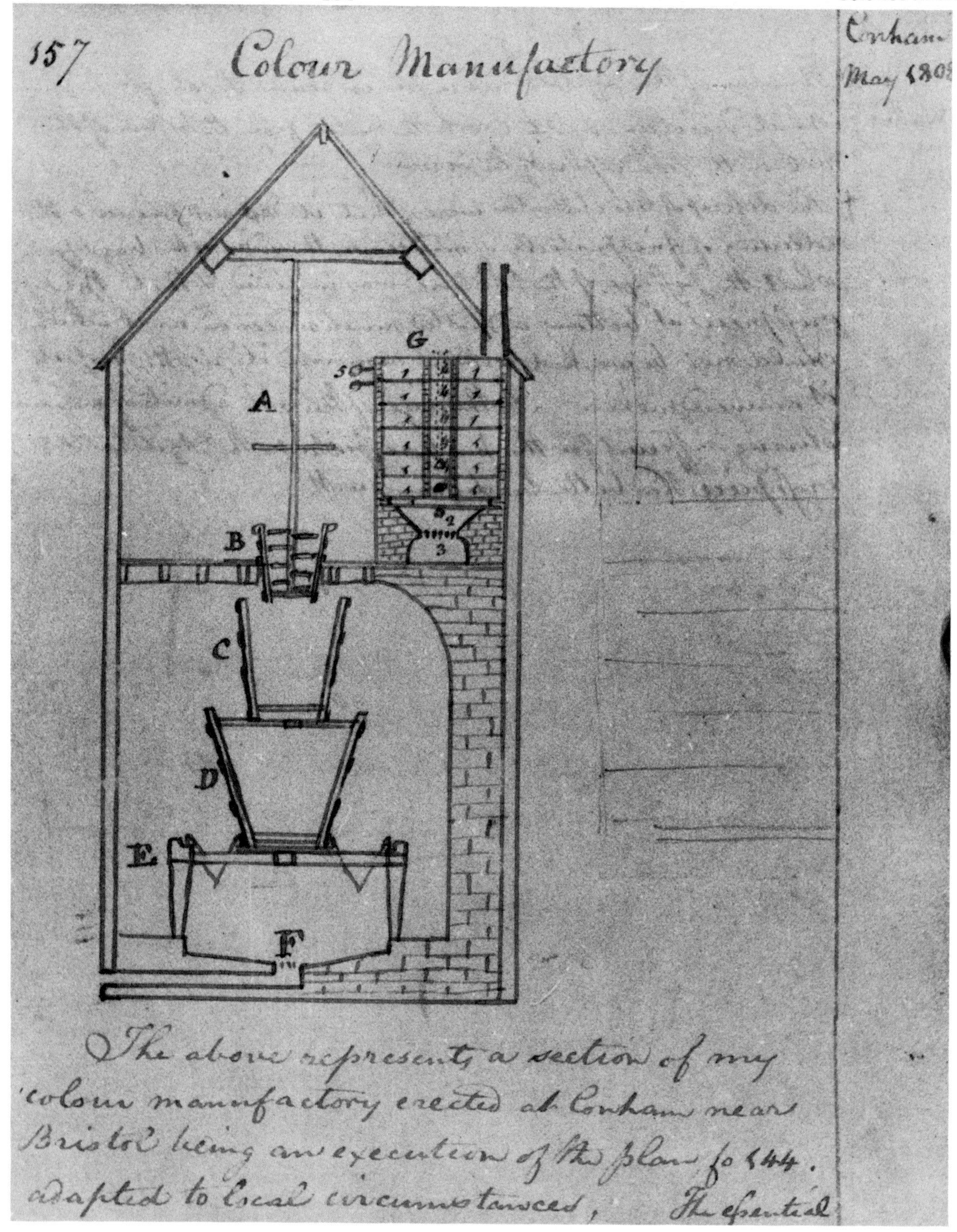

Fig. 52. Field's sketch of his lake-making factory at Conham, 1808. Vats were arranged on different levels so that the dye might be poured on to the base and, after mixing, the contents of the lowest vat D were emptied into conical filters E in which the lake pigment was collected while water ran away at the drain F. (Field, 'Practical Journal 1807', Courtauld Institute of Art MS. Field/3, f.157.) (Reproduced by courtesy of the Courtauld Institute of Art.)

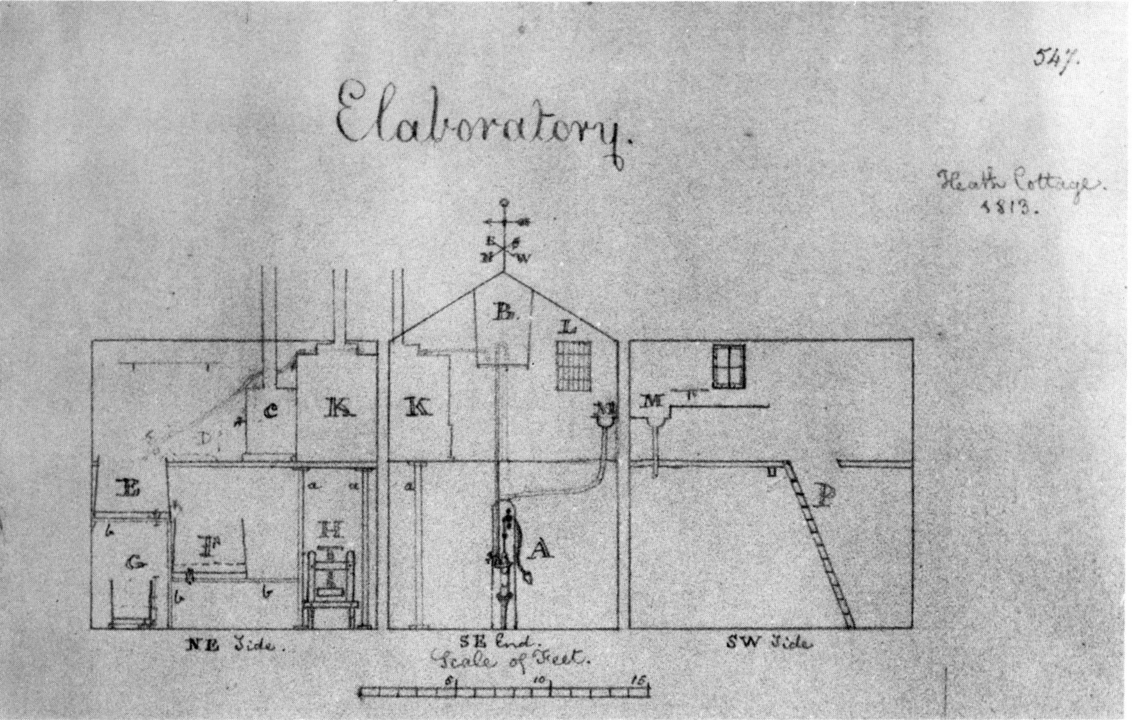

Fig. 53. Field's lake-making factory at Hounslow Heath, 1813. A is a water pump, B a water cistern, K air stoves, C a copper boiler for hot water required for the mashing tub D, from which the hot, liquid dye was strained into tincture tub E, from which the dye was poured on to the substrate in precipitating tub F. When reaction was complete, the contents were poured into the filtration apparatus, physeter G, where the pigment was collected. (Field, 'Practical 1808', Courtauld Institute of Art MS. Field/5, f.547.) (Reproduced by courtesy of the Courtauld Institute of Art.)

date of preparation, which has frequently been ignored by later writers, many of whom erroneously give the year of the report as the date of discovery. Scheele's green is an example, as, almost without exception, writers on pigments state that it was first made in 1778, whereas the report itself suggests that 1775 is much nearer the truth. For evidence concerning the characteristics of new colours as used in oil and water-colour painting, that supplied by scientific papers is less dependable, as it is clear that scientists and painters often had different points of view. Permanence to light is an important requirement for artists' colours, and most of the chemists realised it, but, even so, their tests for permanence were either insufficiently exacting or of too short a duration, with the result that exaggerated claims were made concerning the durability of colours such as chrome yellow and red. It would hardly be surprising if such unfounded claims strengthened the natural caution of some painters. After all, if they hoped that their paintings would last for centuries, it was hardly too much for them to expect the pigments they used to have been in existence and under observation for at least a decade. Then there were other desirable attributes, such as brilliance, tinting power, transparency or opacity, good consistency and drying time, which were seldom discussed extensively in scientific papers. Sometimes a chemist regarded a non-poisonous quality as more important than all other characteristics excepting permanence alone, but painters were not unduly concerned about toxicity. In spite of the difference in viewpoint between scientists and painters, it is fair to add that several scientists were genuinely interested in the practical application of their discoveries, and a few, notably Chaptal and Davy, were sufficiently interested in traditional pigments to analyse some which had been used in classical antiquity. In regard to the shortcomings of new colours as compared with scientists' claims, it is necessary to bear in mind the fact that chemists' reports were based on small laboratory samples; it is possible that a pigment which a chemist regarded as satisfactory under such circumstances, was, through no fault of his, less satisfactory when manufactured commercially on a large scale. For various reasons, therefore, it is desirable to compare the remarks made in scientific papers with the opinion of colourmen who combined a scientific and practical approach.

From the beginning of the nineteenth century, colourmen produced books in which they passed on that part of their knowledge which was useful to artists, thus providing a link between technology and the practice of painting. The books by colour-makers de Massoul and Field provided a broad survey of the pigments and media available in their day, and the shorter works of colourmen Ackermann and Winsor & Newton supplied details of their own water colours. Those books and early catalogues show that a very wide range of colours was supplied; not only had some traditional pigments been revived in the nineteenth century, but different varieties of one pigment were often sold under several names. The correct interpretation of colour names at all periods presents a problem, because a tendency to refer to relatively new colours by

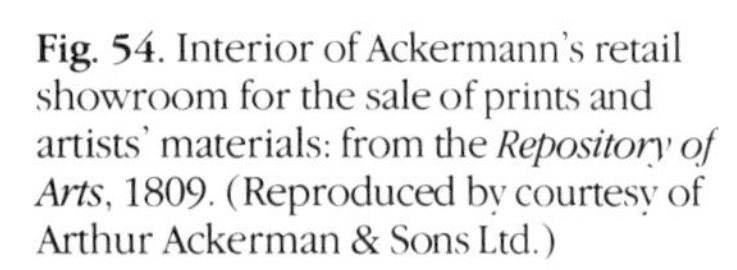
Fig. 54. Interior of Ackermann's retail showroom for the sale of prints and artists' materials: from the *Repository of Arts*, 1809. (Reproduced by courtesy of Arthur Ackerman & Sons Ltd.)

traditional names has always existed in order to give artists some idea of the character of a colour. Thus, bice was composed of smalt instead of azurite in the eighteenth century, gallstone was a quercitron lake pigment in the nineteenth century, and similar examples are to be found amongst twentieth-century colour lists. Great care is needed in the interpretation of colour names, and it is, unfortunately, not always evident in present-day papers on old pigments where lists of synonyms and obsolete terminology are included without reference to date or documents. However, comparison of documentary sources supplies the meaning of most colour names. For example, colourmen did not always explain the composition of every colour on sale in the early nineteenth century, but it is possible to identify many of them through reference to associated literature. Ackermann's book of 1801 may usefully be compared with that by de Massoul. The composition of Field's lemon yellow is established by reference to Bachhoffner; the composition of some of Ackermann's colours *circa* 1840 is mentioned by Fielding, and some of Winsor & Newton's colour names can be interpreted in the light of information provided in the book by J. S. Taylor, the firm's research chemist in the late nineteenth century.

In the past colourmen's literature doubtless helped artists to select colours from amongst the wide range, and, at the same time, it raised the prestige of the British colour industry. The volume of trade had expanded enormously during the eighteenth century, a fact which is clearly attested by the evidence of customs records showing export of increasing quantities of 'colours for painters' to many parts of Europe and America throughout the period. However, according to several artists, the integrity and knowledge of some colourmen was questionable. At the end of the eighteenth century there was a plea for the Royal Academy to pass judgement on the quality of artists' colours and media, and a little later an Academician suggested that an official body should be instituted for the preparation of colours and other equipment such as canvases.[346] While the complaints may have been justified to some extent, it is only fair to state that some painters laboured under the delusion that they lacked superior materials which had been used by the old masters when in fact what they required was a thorough knowledge of their craft. Evidence of a growing feeling that a groundwork of technical knowledge was desirable is reflected in a letter to *The Artist* in 1810, which contained the statement: 'chemistry is to painting what anatomy is to drawing. The artist should be acquainted with them, but not bestow too much time on either.'[347] Colourmen's literature could not entirely fulfil the needs of artists for a knowledge of chemistry, but at least it supplied details of the characteristics of pigments and media so that there was no need for a painter to be in complete ignorance of his materials. Field was quite outspoken about the fact that the ultimate responsibility for the choice and use of artists' colours belonged entirely to artists themselves:[348]

> It is due to the respectable colourmen of the present day to bear testimony to the laudable anxiety and emulation with which they

> purvey, regardless of necessary expense, the choicest and most perfect materials for the painter's use; so that the odium of employing bad articles attaches to the artist, if he resort to vicious sources or employ his means improperly.

Field was a firm believer in the precept that art and science are indivisible and that an artist's work is better for his having full understanding of 'the philosophy of his art'. Three centuries earlier painters could well have been surprised that the point needed such strong emphasis, because, to them, science as applied to anatomy, perspective and paint was an integral part of their craft. Yet, as the social status of painters had changed and artists were no longer classed as artisans, they dissociated themselves from much that was practical or mechanical and concentrated on the aesthetic and intellectual aspect of fine art. Many painters did in fact retain a degree of craftsmanship, and they were practical in that they actually practised painting; nevertheless, documents on painting of the eighteenth century reflect a degree of separatism, showing painters as a class above the artisans who prepared their colours and on a plane apart from the chemists who conducted the scientific research which resulted in the discovery of new pigments. However, as the industrial revolution progressed in the early nineteenth century, painters became increasingly aware that art and science should be complementary. The report of a committee set up in 1835 to examine the connection of the arts with manufactures shows quite clearly that artists wished to identify themselves with the new industrial world; it reflects dissatisfaction with the limitations of art education as it existed then, limited as it was to fine art for a few, at a time when designers were needed in industry and a far wider section of society was ready to benefit from both academic and technical art education. The findings of the committee of 1835 helped to bring about the establishment of the schools of design a few years later with the purpose of gaining closer collaboration between art and industry.[349] The object was not altogether easy to achieve, although technical teaching as a part of art studies was accepted in principle and from there it was a relatively short step towards greater concentration on the technology of painting. Study of the technology of painting materials meant a revival in craftsmanship, not the practice of methods supposedly based on resurrected formulae of past ages, but the pursuance of sound technique on the basis of scientific principles.

Literature on pigments and painting reflects various changes between 1600 and 1835. In the early seventeenth century the workshop tradition was still strong, and many painters were able to acquire sound knowledge of their materials, although, even then, some work in the preparation of colours was being undertaken by artists' colourmen. With a few exceptions, books written during the eighteenth century reflect a general lack of knowledge concerning the technology of colours; eighteenth-century painters were at a disadvantage in that they had little information with which to supplement their studio or academic training. During the early nineteenth century the range of pigments was wider than ever before, owing to the contribution of chemists and colour-makers, yet the necessity for painters to make wise use of their colours was correspondingly great. Fortunately, increased availability of technological information together with a new attitude to art training gave painters in the nineteenth century a considerable advantage over their immediate predecessors, and they were consequently in a position to make intelligent use of the contribution of science to art.

Appendix One

Abbreviations

Publications

Conservation	International Institute for Conservation, *Studies in Conservation*
O.E.D.	*Oxford English Dictionary*
Phil. Trans.	Royal Society of London, *Philosophical Transactions*
P.M.L.A.	*Publications of the Modern Language Association of America*
P.P.V.	*Peintures, Pigments, Vernis*
Walpole Soc.	*Walpole Society Publications*

Institutions

B.M.	British Museum
I.O.R.	India Office Records, Foreign and Commonwealth Office
P.R.O.	Public Record Office
V. & A.	Victoria & Albert Museum

Public Record Office References

C.66 — Records of the Chancery, Patent Rolls
Example: C.66/1682/49=Patent Rolls, 3 James I, part 20, membrane 49

E.190 — Records of the Exchequer, Port Books
Example: E.190/34=Imports to the Port of London, 1630

SP.14 — Records of the Secretaries of State
Example: SP.14/22/30=State Papers Domestic, James I, 1606, volume 22, number 30

Customs — Records of the Board of Customs and Excise
The selection of ledgers of imports examined was as follows:

3/4	1700	3/80	1780
3/13	1710	5/1A	1792
3/22	1720	5/1B	1800
3/30	1730	4/6	1810
3/40	1740	4/9	1814
3/50	1750	4/15	1820
3/60	1760	4/25	1830
3/70	1770		

Appendix Two

Books with Named Colour Samples

Comments on the condition of the colours in books containing named samples are relevant only to the particular volumes examined, and, for that reason, details of those consulted are given below. Some traditional earth colours are included in all the books, so only the more unusual or interesting colours are mentioned here. The volumes are listed in chronological order.

British Museum, Manuscript Room: MS. Sloane 2052.

De Mayerne, 'Pictoria, Sculptoria et quae subalternarum artium' (*c.* 1620–40), ff. 80–82v.

The de Mayerne manuscript contains four pages of samples in water colours and also gold and silver. The text is in neither Mayerne's handwriting nor his language (the colour names are in Latin and German) but the pages are nevertheless contemporary with the rest of the manuscript. Some of the colour samples are in poor condition because they are badly rubbed, while others have undergone colour alteration.

Folio 80 contains examples of two white colours, lead white and chalk, together with examples of gold and silver.

Folio 80 *verso* is reproduced in *Plate 7*. The samples fall into four groups:

1. The upper two rows contain ten samples of five black and brown colours. The sample at top right is Cologne earth.
2. Yellow samples are included in the third and fourth rows. From left to right in the third row are: saffron which is shown as an extremely transparent, rather pale yellow; massicot that has darkened completely; light and dark ochre. The fourth row has four samples of two varieties of orpiment and four samples of two varieties of organic brown-yellows of the type that used to be called *pink* in English.
3. Red samples are included in the fifth and sixth rows. From left to right they are: minium (red lead) that has blackened, vermilion, Brabant lake and Paris red. The sixth row contains a red-brown earth, four examples of two varieties of a colour that seems likely to be based on brasil, and Venice lake.

4. Green samples are included in the two bottom rows. They include two varieties of copper green, chrysocolla (*Berggrün*)—a naturally occurring mineral green—and *Lasurgrün* that may also come from a naturally occurring mineral. Beneath these bright greens are two varieties of sap green that have discoloured badly.

Folio 81 *recto* is reproduced in *Plate 8*. The samples fall into two groups.

1. The upper two rows contain samples of blue water colours. The first row has, from left to right, six samples of three degrees of azure blue. The fact that they are shown in three grades suggests that the colour is ultramarine, but alternatively it may be azurite. At top right is a deep variety of smalt, while a paler variety, much rubbed, is the first colour in the second row. This is followed by indigo, litmus and turnsole. The last two have reverted from blue to red or lilac.
2. The remainder of the page is occupied with examples of mixed purples made up from, upper row, left to right, the three grades of azure blue, two grades of smalt, indigo and litmus (in the second and succeeding rows the places of indigo and litmus are reversed), each of these mixed with, from top to bottom, red lead, vermilion, Paris red, Brabant lake, and brown red. Generally, the mixtures containing either Paris red or Brabant lake are the most successful.

Folio 81 *verso* contains examples of mixed green water colours.
They include saffron, lead-based yellow, light ochre, brown ochre and two organic yellows, each mixed with copper green, chrysocolla (*Berggrün*), azure, and two varieties of sap green. The mixtures with copper green form the strongest green.

University of London Library: PR [N-Royal].

R. Waller, 'A catalogue of simple and mixt colours', *Phil. Trans.*, xvi (1686), 24–32.

Includes small samples of twenty-one hues and a large number of mixtures. The main colours include Spanish white, azurite, ultramarine, smalt, litmus, indigo, Indian ink, ceruse, massicot, gamboge, orpiment, red lead, vermilion, carmine, lake and dragon's blood. Apart from Spanish white (indistinguishable), smalt (very pale), massicot and red lead (black), the samples are in good condition. The names are given in Latin.

(Only the first edition contains these samples. They are not included in a later reprint containing volume xvi and others that is catalogued under the same press mark in the University of London Library.)

British Museum, Print Room: 167.a.8.

J. C. Ibbetson, *Process of Tinted Drawing*, n.d. [1794].

Samples of nine hues, three dilutions of black ink, and three sky tints are included. The hues comprise gamboge, yellow ochre, raw and burnt sienna, light red, lake, Prussian blue, brown pink and Vandyke brown, all in good condition.

Victoria & Albert Museum: Box III.107.G.

J. C. Ibbetson, *An Accidence or Gamut of Painting* (1803).

Ten hues and ten mixtures. The hues comprise flake white, Naples yellow, raw and burnt sienna, vermilion, light red, lake, Prussian blue, brown pink and Vandyke brown. The white has deteriorated.

British Library: 797.dd.24.

J. Roberts, *Introductory Lessons* (1809).

Plate 6 contains samples of thirteen hues and eight mixtures. Hues include orange orpiment (realgar), gamboge, red lead, lake, Prussian blue, indigo, Vandyke brown, Cologne earth, and Indian ink, as well as ochre and sienna.

British Museum, Print Room: C.167*b.36.

J. Roberts, *Introductory Lessons,* another copy.

It lacks samples of realgar, burnt sienna, Cologne earth and all but one of the mixtures.

Courtauld Institute of Art, University of London: MS. Field/6.

George Field. 'Examples and Anecdotes of Pigments. Practical Journal 1809'.

Contains about 276 water-colour samples, some 100 of which were painted out and notes regarding performance in fading tests written up in 1809. The remaining samples were added later. The manuscript is a very useful source for pigments available in the early nineteenth century and it is cited as Field's 'Practical Journal 1809' in the preceding text when reference is made to particular colour samples.

Victoria & Albert Museum: 65.E.3.

J. Varley, *J. Varley's List of Colours* (1816); bound with *A Treatise on the Principles of Landscape Design* by the same author.

The list includes spaces for nineteen hues and mixtures. The space for cobalt blue is blank, but colours which have been inserted are Prussian blue, indigo, lake, gamboge, burnt sienna, yellow ochre, Venetian red, vermilion, burnt umber, Roman ochre and sepia.

Science Reference Library (Kean Street Annexe), part of the Reference Division of the British Library: BQ*50, accession number 5552.

Anon. *Le maître de miniature* (1820).

Twenty-one hues and twenty mixtures. The hues include carmine, lake, vermilion, red lead, Prussian blue, *bleu minéral* (copper oxide), cobalt blue, chrome yellow, gamboge, yellow berries, sap green, Cologne earth, bistre, Chinese ink, ivory black and various ochres. The samples of cobalt blue and red lead are in poor condition.

British Library: 561.b.22(2).

T. H. Fielding, *Index of Colours and Mixed Tints* (1830).

This includes several pages with samples of twenty-eight colours and many mixtures. Amongst them are chrome yellow, gallstone, Indian yellow, Italian pink, madder lake, cobalt blue, Antwerp blue, brown madder, sepia, brown pink and emerald green. All the samples are in excellent condition.

Victoria & Albert Museum: 41.D.50.

J. Scott Taylor, *A descriptive Handbook of modern Water-Colour Pigments,* n.d. The book was written in 1887 but reprinted many times; this volume is from the fourteenth thousand.

It contains four pages with a total of ninety-six colour samples. They include several varieties of madder lake, vermilion, ultramarine and sepia, cadmiums, chromes and Mars colours, oxide of chromium, viridian and emerald green, and quercitron lake under various names such as yellow lake and gallstone.

Appendix Three

Patents for Colour-Making in the Early Seventeenth Century

During the first half of the seventeenth century, industrial activity in England increased to such an extent that it might almost be regarded as a minor industrial revolution. There are several reasons why it never reached the same proportions as the eighteenth-century revolution, but the experience of certain colour-makers indicates that government measures which were intended to encourage manufacturers provided instead an effective bar to industrial enterprise.

In order to encourage native industry, patents for invention were granted to successful applicants regardless of the existence of the same industry overseas; all that mattered was that the trade should be new to England. This was emphasised in the letters patent for smalt-making issued to William Twynyho, Abraham Baker and John Artogh in 1605, the earliest English patent for making a pigment. According to the grant, the patentees were given sole right for twenty-one years to make the blue colour which was described as follows:[350]

> a certayne blewe stuff called Smault, commonly used by Paynters and Lymmers, whiche hathe not at any tyme heretofore bene made . . . within our said realme of England . . . and which shalbe as good, perfect and merchantable as the same or like stuff called Smaulte made, wroughte and compounded in the partes beyonde the seas.

Import of the pigment or the materials for making it by anyone other than the patentees was prohibited, and they were granted the customary right of search which allowed them to inspect the premises of suspected competitors. Another clause provided that the maximum selling price of their smalt should be no higher than that current during the seven years preceding the date of the grant.

The industry was in jeopardy from the start, for the Dutch took prompt action to oppose the grant and avert the loss of a valuable market. The manufacturers at Middleburg, then capital of the province of Holland and an important glass-making centre, were the instigators of opposition. Their moves are best described in the words of Twynyho who appealed for the maintenance of the patent in July 1606, complaining that the foreigners had obstructed him:[351]

1 By sending over an extraordinarie quantitie [of smalt] and at a base price.
2 By sending over two Straungers with Mill and Stocke to worke here.
3 By procuringe certaine grocers to exclaime against us in parliament, alleadginge Ibie:
1 That we trouball them in searchinge theire houses.
2 That we could make none good.
3 That we inhannsed the price.
4 That yt hath bin made heretofore in England.

In answer, the patentee stated that only four houses had been searched, each with the owner's consent, and that people who had opposed the grant earlier were now buying the patentees' smalt and commending it. The beginning of financial difficulties is evident in the reply to the third charge and in the details of the initial expenditure in starting the industry:[352]

3 Yt is provided by our Patent, that we shall not sell above the price, that hath bin for the most part of 7 yeares last past and nowe we sell farre under compelled thereto by the grate quantitie formerlie sent into the Realme to over throwe us. We sell for x^{d} the lb. which is the verrie rate in the Custome booke, to our present losse.
4 There is requyred for the makinge of this Smault many and greate Romes, dyvers milles and a great stocke, to the vallue of 2000^{li}. Yf any can shewe that this hath bin donne here we will reiect our Patent.

Twynyho also tried to correct an erroneous idea, apparently held by some Members of Parliament, that smalt was made from wheat and was therefore prejudicial to the national interest. (Confusion had arisen because a pale variety of smalt called blue starch was used by laundresses as a washing blue, and it was consequently identified with starch used for stiffening linen.) Twynyho emphasised that it was made from imported commodities and that the patentees were contributing to the revenue by paying £244 annual customs and rent. His appeal concluded with a claim that the new industry should be encouraged as it was a source of employment. The last point carried some weight, as, in a report concerning grievances about grants which was read to the Commons in November 1606, it was allowed that the smalt industry would be 'a means to set many poor People a-work'.[353]

The original grant was cancelled in February 1609 and replaced by another in which Abraham Baker alone was named as patentee with sole right to organise smalt manufacture for thirty-one years.[354] This time he was allowed to use a seal, bearing a half-lion holding a sceptre in one paw and a posy, so that his smalt should be distinguishable from unauthorised supplies. The stipulation about the selling price remained the same, but Baker was to pay a lump sum of £240 annually instead of paying £24 rent and £220 customs as before. There is no indication of the reason for the withdrawal of Twynyho and Artogh, but the name of another appears near the end of the grant in a stipulation that Baker should appoint deputies only with 'full consent and good likinge of Sir George Hay,

Knighte, one of the Gentlemen of his Maiesties privy Chamber', thus indicating that Baker was dependent on the influence and possibly the financial support of someone close to the king, the individual named being a member of the Scottish entourage of James I which arrived with him in 1603.

It was extremely difficult to enforce the regulations concerning industrial monopolies, and during the following years competitive smalt manufacture existed in London. It will be remembered that Twynyho had complained that the Dutch had sent over men and equipment to establish the industry in an attempt to invalidate the first grant, but he never said what had become of them. Later evidence concerning a Dutch-born but naturalised English manufacturer suggests that he was the immigrant referred to, that he may well have introduced the industry to England before the first patentees and that he was still in business in 1613.

Christian Wilhelm, a free denizen, ran a smalt-making concern at a house called Pickleherring on the south bank of the Thames in St Olave's parish, Southwark, in a district which was noted in the seventeenth century and later for glass-making and on a site which can be determined exactly by Pickle Herring Stairs and the existing Pickle Herring Street. When prosecuted by Baker, Wilhelm was able to supply documents in support of his claim to have started manufacture in 1604. A certificate signed by customs officers testified to the fact that he had imported saffer and millstones in that year and that he had exported smalt at the beginning of 1605. A letter from his factor and another from his landlord, the owner of Pickleherring, confirmed that Wilhelm had set up smalt manufacture in 1604. Several blacksmiths detailed money paid to them for setting up the mills and shoeing the horses; the latter were particularly fine and four of them had been requisitioned by the king. Grocers of the City of London, who had purchased the pigment, and parishioners of St Olave's supported Wilhelm's claim, and several accused Baker of obtaining knowledge of the industry from Wilhelm's workmen so that he could then set himself up with a claim to be the first manufacturer:[355]

> Att his firste Comynge hether he the saide Christian Wilhelme was reputed and knowne to be onelie skillfull in the makinge and Compoundinge of a certaine blew stuffe named Smalte, but commonly called blewe starch, and brought the mystery thereof with him into this Realme. And that shortlie after his beinge here, and after his exceedinge charge of takinge a howse, erectinge of Ovens, Milles and other necessaries, William Twyniho, Abraham Baker, and John Artogh secretlie enticed one or some of the saide Christian his servauntes, whoe (as it is enformed and confessed) gave them some lighte in the saide arte. And so by means . . . procured his highnesses graunte . . . wheareby the said Christian notwithstanding that he lefte his owne Contrie, broughte the saide arte and mystrie first hither, and here hathe adventured his whole estate in the charge thereaboutes, is by the meanes aforesaide deprived of the benyfytt thereof, to the undoinge of him, his wife and famylie.

The Painter–Stainers Company offered a testimonial to the good quality of Wilhelm's smalt (which he had requested), but went further, accusing Baker of serving them badly so that they had been forced to buy foreign smalt, and ending with a plea for free enterprise in industry:[356]

> We thincke in our opynions that the more Smalte makers permitted will not onelie be proffytt to his Majestie in regarde of his higheness subjectes that mighte be ymployed thereaboutes. And the Company better served and better cheape.

The documents from which these details come are copies made in the Secretary of State's office from originals which were presumably submitted in 1613, as some bear that date. No details of the actual court proceedings have been found, and it seems likely that evidence concerning the case disappeared with other records of Privy Council proceedings, 1603–13, which were lost in the fire at Whitehall in 1619. The fact that the testimony of the customs officers, the factor and the landlord was dated 1607 suggests that Wilhelm had been required to explain his position on a previous occasion, and it lends a certain amount of credence to his claim. Nevertheless, later documents make it clear that the patentee's case was upheld, and one must suppose that all Wilhelm's stock was confiscated and his equipment—the mills, vessels and ovens mentioned in many of the depositions—was destroyed as directed by the provisions of the patent. Such seizure and destruction was one of the major grievances concerning monopolies; it gave rise to Section IV of the Statute of 1624, whereby an injured party was allowed to sue for relief at common law in order to receive damages amounting to treble the value of his confiscated property. Legal protection came too late for Wilhelm, but he was not entirely ruined, for he is mentioned in later records (compiled as a result of popular antagonism to immigrant tradesmen) as a vinegar distiller and gallipot-maker resident in the Maze on the Waterside at Southwark.[357]

As for Baker, the elimination of a competitor still could not guarantee prosperity. Smalt of foreign origin was still imported, so that in 1618 letters were issued to mayors and officers of the Cinque ports instructing them to prevent the illegal entry of the pigment.[358] The patentee was in such financial difficulty that the 1609 patent was cancelled, and another was issued to Baker in 1619 in which he was required to pay only £20 per annum and no mention was made of annual rent. In effect, Baker probably paid nothing at all because the customs payable to the Crown on monopoly smalt had been transferred to Sir George Hay.[359] The adjustment afforded only temporary respite; the illicit import and sale of smalt continued, and in 1623 Baker was bankrupt. 'The foul and unconsionnable practices have bin such as have brought the poore man to utter ruine,' wrote Lord Conway in support of Baker's suit for relief *in forma pauperis.*[360] The monopoly was allowed to continue, for it was one of the few existing grants mentioned in the Statute of Monopolies (Section XIV) as being exempt from its provisions, but parliamentary recognition was of no avail. According to the terms of the patent, no smalt should have been imported by anyone other than Baker until 1650, but an examination of a London port book for 1630 shows that quantities of smalt amounting to many tons were imported openly by various people.[361] The collapse of the native industry must be attributed wholly to the monopoly system. No one but the patentee was permitted to practise the trade, but the patentee was not in a position to run a successful business owing to the price restrictions contained in the grants. By the patents of 1605 and 1609 the price was to be no higher than that current in the seven-year period 1598–1605, and by that of 1619 it was to be no higher than during the period 1602–9. The restriction was well-intentioned and designed to prevent profiteering, but it acted most unfavourably during a period of inflation and it meant that the patentee could not make a living, still less supply sufficient smalt for the needs of the whole country.

By comparison with the twenty-year struggle to establish the smalt industry, the patent for preparing indigo was short-lived. The East India Company had a virtual monopoly of all indigo sold in England at the beginning of the seventeenth century, but during the 1630s the Company had to face competition and an alarming drop in the value of their indigo (which they had hitherto sold for about six shillings and sixpence per pound), owing to the sale of a particularly good-quality product from the West Indies and, also, to the sale of refined indigo which was treated by an industrial process in England. In February 1634 the govenor of the East India Company mentioned the latter, which he described as an abuse, saying that 'in making a deceiptfull and counterfait Indico, in and about London . . . the parties that nowe make the flatt Indico are growne to that excellent Art and Cunning as neither by the coullor nor yet by the breaking of the Indico, the falsity can be discerned from that which is real and good.'[362] From his condemnation one might think that the competitive product was some kind of imitation, but it was actually genuine indigo, skilfully prepared, as one may learn from the patent which was granted to William Bolton, a grocer, in July 1634.[363] Two processes were patented at one time: by one, Bolton removed sand and dirt from flatt indigo (that is, dust of indigo, which was often contaminated), separating the pigment so that it was as good as the best quality, rich indigo; by the other process, he ground and soaked lumps of rich indigo so that they were easier to use. The grant supplies no details of the processes employed, but they were undoubtedly similar to the washing over process commonly employed by painters who wished to separate fine pigment particles from coarse particles or dirt. In any case, the patentee was almost certainly the same William Bolton who had been granted letters patent for refining red ochre eight years earlier. In the earlier grant, mills, vessels and drying ovens are specifically mentioned.[364] To undertake washing on an industrial scale, a series of tanks on different levels would have been required. It is disappointing that Bolton provided no details of his equipment, but reports of the quantities of refined indigo which he supplied for the home market and for export prove that he conducted operations on a large scale.

It must have been particularly galling for members of the East India

Company to see someone making a profit out of dust of indigo, which they often referred to as useless and generally sold at a low price, about one-sixth that of rich indigo, and which in 1619 they had even considered giving away. Bolton was well able to buy the dust from them, treat it, and sell the refined product more cheaply than rich indigo while still making a good profit. All the Company did to their indigo was to garble it by sieving, and even that operation was not efficiently carried out at one time, for in 1619 the Company officials found that the workman was using a worn sieve which allowed pieces of rich indigo to pass through with the dust.[365] There is no reason to suppose that they had improved their method by the 1630s, because they did not challenge the validity of Bolton's claim to have introduced an industrial process which had not been practised before; instead, they attempted to prove that Bolton's indigo was inferior and that he was deceiving the public. After prevailing upon various dyers to certify that Bolton's indigo was bad, they resolved to appeal to the Attorney General on the grounds that the patent was 'contrary to Lawe, mischievous to the State and prejudicial to the Company'.[366] The charges sounded convincing, because the Company heard in December that the patent had been withdrawn pending an enquiry by the mayor and four aldermen of London who would take evidence from the dyers.[367] Bolton, who apparently wished to avoid court proceedings, offered to surrender his patent voluntarily and to give an undertaking to discontinue his industry if the Company would take back his stock of indigo at the same price as he had paid. Its value amounted to £6000, but the Company refused his offer on the grounds that 'they were noe buyers but sellers of that Comodity and for them to ease him and preiudice themselves is noe reasonable mocion'.[368] Practical tests of Bolton's indigo and that supplied by the Company dragged on during the succeeding months, as, when the Company had difficulty in obtaining sufficiently condemnatory reports concerning Bolton's indigo from some dyers, they promptly enlisted others to make further trials. This suggests that the Company had insufficient grounds for complaint, but they continued to wear down the patentee and in August 1635 they refused another offer of surrender from Bolton (whose industry and export trade still continued) and they decided to let their charges of fraud stand for submission to the mayor and aldermen.[369] No evidence of a court decision has been found, but Bolton is not mentioned in subsequent records of the Company, so it is likely that he gave up refining indigo. It is evident that he was sufficiently harassed to feel that his patent afforded little protection and that he might as well be without it. The East India Company was sufficiently powerful to hinder an individual, but it could not halt industrial activity entirely. When the officials received an offer for dust of indigo in 1640, they sought an undertaking that the buyers would not treat it and make it into solid lumps of good pigment, but the potential buyers refused to give an assurance and the dust of indigo was put up to auction in the usual way.[370]

An unfortunate feature of documentary evidence concerning early patents is that it concerns manufacturers who were in trouble. The early seventeenth century was undoubtedly a difficult time to develop large-scale industrial activity, owing to inflation and government interference, and even minor concerns could suffer badly as a result of pressure from vested interests and anti-foreign agitation. Nevertheless, an account of two unsuccessful concerns shows that considerable capital was invested in industry and different branches of colour-making were practised on an industrial scale. Painters may still have made a few pigments for themselves, but by the seventeenth century it was the exception rather than the rule.

Bibliography

Manuscripts

British Museum, London

Egerton 1636.
Harley 1279, 6000, 6376.
Sloane 122, 228, 288, 1394, 1448B, 1990, 2052, 3292, 6284.
Stowe 680.
Additional 4459, 6032, 12461, 23080, 34120.

Berger, Jenson & Nicholson, London

A collection of records dating from the late eighteenth and nineteenth centuries, comprising manufacturing formulae and delivery, invoice and stock records. They are unnumbered and in the main untitled, so the descriptions used by Bristow have been adopted here.
Lake and Carmine Book.
1831 formula book.
First, second and third experimental notebooks.
Stock list and accounts for 1801, 1805 and 1809.

Courtauld Insitute of Art, University of London

Field 1–Field 10. Ten manuscript notebooks by George Field.
Field 6. 'Examples and Anecdotes of Pigments. Practical Journal 1809' is referred to in the text by the abbreviated title 'Practical Journal 1809'.

Royal Society of London

Manuscript 136.

Sir John Soane's Museum, London

AL. 41 C.

Victoria & Albert Museum, London

86.EE.69 Formerly 86.L.65 as cited in the first edition of *Artists' Pigments. c. 1600–1835.*
86.FF.19 Formerly 86.H.16.
R.C.A.20 The manuscript cited under this press mark in 1970 is now officially listed as missing.

Edinburgh University Library

Laing III 174 (photocopy in the Victoria & Albert Museum).

Bodleian Library, Oxford

Ashmole 768, 1399, 1480, 1491, 1494.
Rawlinson D.1361.
Tanner 326.

Manchester Reference Library

923.9 D55.

Periodical Publications

Annales de chimie et de physique, Paris, 1st-8th series 1789–1913.
Annals of Medical History, New York, 1st-3rd series 1917–42.
Annals of Philosophy, London, 1st series 1813–20, 2nd series 1821–26.
Annals of Science, London, 1936, in progress.
Burlington Magazine, London, 1903, in progress.
Chemistry and Industry, London, 1918, in progress.
CIBA Review, Basle, 1937, in progress.
Deutsche Akademie der Wissenschaften zu Berlin, *Miscellanea Berolinensia,* Berlin, 1710, 1723.
International Institute for Conservation, *I.I.C. Abstracts: abstracts in the technical literature on archaeology and the fine arts,* London, 1955, in progress.
International Institute for Conservation, *Studies in Conservation,* London, 1952, in progress.
Journal des mines, Paris, 1794–1815.
Journal of Natural Philosophy, Chemistry and the Arts, ed. W. Nicholson, London, 1797–1813. Often referred to as *Nicholson's Journal.*
National Gallery Technical Bulletin, London, 1977, in progress.
Peintures, Pigments, Vernis, Paris, 1924, in progress.
Royal Society, *Philosophical Transactions,* London, 1665, in progress.
Society for the Encouragement of Arts, Manufactures and Commerce, later the Royal Society of Arts, *Transactions,* London, 1783–1849.
Technical Studies in the Field of the Fine Arts, journal of the Fogg Museum, Harvard University, published at Lancaster, Pa., 1932–42.
Walpole Society [*Publications*], Oxford, 1912, in progress.

Printed Books and Papers; University Theses

The location of certain rare works is inserted in brackets after the biliographical details of each one.

Academia Italica, the Publick School of Drawing, or the Gentlemans Accomplishment, London, 1666.
Ackermann, Rudolph, *A Treatise on Ackermann's Superfine Water Colours,* London, n.d. [1801]. ßBirmingham Reference Library, photocopy in the University of London Library.]
Ackermann, Rudolph, *The Repository of Arts, Literature, Commerce, Manufactures, Fashions, and Politics*, London, 1809.
Ackermann's Manual, see Fielding.
Agricola, Georgius [Georg Bauer], *De re metallica,* translated and eds. H. C. and L. H. Hoover, London, 1912.
Alston, J. W., *Hints to young Practitioners in the Study of Landscape Painting,* London, n.d. [*c.*1804].

Armenini, Giovanni Battista, *De veri precetti della Pittura,* Ravenna, 1586.
The Art of Drawing and Painting in Water-Colours, London, 1731; further editions and reprints, 1732, 1755, 1757, 1770.
The Art of Painting in Miniature: teaching the speedy and perfect Acquisition of that Art without a Master, London, 1729. Translated from French, see C.B. Reprinted in 1730 and in *Arts Companion,* 1749.
The Art of Painting in Water-Colours: exemplified in Landscapes, Flowers, &c., London, 8th edn, 1786; 18th edn, 1818. See Bowles.
The Artist: A Collection of Essays, relative to Painting, Poetry, Sculpture, Architecture, the Drama, Discoveries of Science, and various other subjects, 2 vols, London, 1810.
The Artist's Assistant in Drawing, Perspective, Etching, Engraving, Mezzotinto-Scraping, Painting on Glass, in Crayons, in Water-Colours, and on Silks and Satins, London, 5th edn, 1788.
The Artist's Assistant, in the Study and Practice of Mechanical Sciences, privately printed, n.d. [*c.*1790?].
The Artist's Assistant: or School of Science, London, 1807.
The Artist's Repository and Drawing Magazine, 4 vols, London, n.d. First published 1784–86. Vol. ii was later reprinted separately. See *Compendium.*
Arts Companion or a new Assistant for the Ingenious, Dublin, 1749. The book contains reprints of the two preceding works and *The Art of Drawing in Perspective* by the same anonymous author and translator.
C.B. [Claude Boutet?], *Traité de mignature, pour apprendre aisément à peindre sans maître,* Paris, 2nd edn, 1674; further editions 1684; La Haye, 1688; Paris, 1696, 1711; and as *École de la miniature,* 1817.
Bachhoffner, George H., *Chemistry as applied to the Fine Arts,* London, 1837.
Bancroft, Edward, *Facts and Observations, briefly stated, in support of an intended Application to Parliament,* privately printed, n.d. [1798:.
Bancroft, Edward, *Experimental Researches concerning the Philosophy of Permanent Colours,* 2 vols, London, 1813.
Bardwell, Thomas, *The Practice of Painting and Perspective made easy,* privately printed, 1756.
Barret, George, *The Theory and Practice of Water Colour Painting,* London, 1840.
Bate, John, *The Mysteryes of Nature and Art,* London, 1634; 2nd edn, 1635; 3rd edn, 1654.
Beal, Mary R. S., 'A study of Richard Symonds: his Italian notebooks and their relevance to seventeenth-century painting techniques', unpublished Ph.D. thesis, University of London, 1978.
Bearn, J. Gauld, *The Chemistry of Paints, Pigments and Varnishes,* London, 1923.
Beckmann, John, *A History of Inventions and Discoveries,* translated W. Johnston, 4 vols, London, 2nd edn, 1814.
Berger, Ernst, *Beiträge zur Entwicklungsgeschichte der Maltechnik,* 4 vols, Munich, 1901–12.
Birch, Thomas, *The History of the Royal Society,* 4 vols, London, 1756–57.
Boltz von Rufach, Valentin, *Illuminirbuch, künstlich alle Farben zumachen vnnd bereyten,* 1566; Frankfurt, 1589, 1613; Hamburg, 1645; Erffurdt, 1661.
A Book of Drawing, Limning, Washing or Colouring of Maps and Prints and the Art of Painting, with the Names and Mixtures of Colours used by the Picture-Drawers, London, 1666.
Borghini, Raffaello, *Il Riposo,* Florence, 1584.
Bouvier, M. P. L., *Manuel des jeunes artistes et amateurs en peinture,* Paris and Strasbourg, 1829; reprinted 1832 and 1844.
Bowles, Carington, *Bowles's Art of Painting in Water-Colours,* London, 6th edn, 1783; reprinted anonymously; 8th edn, 1786; 18th edn, 1818.
Bowles, Carington, *Bowles's Florist: containing sixty Plates of beautiful Flowers . . . with Instructions for Drawing and Painting them according to Nature,* London, n.d.
Boyle, Robert, *Experiments and Considerations touching Colours,* London, 1664.

Boyle, Robert, *The Works of the Hon. Robert Boyle,* ed. T. Birch, 5 vols, London, 1744.

Bristow, Ian, 'Interior house-painting from the Restoration to the Regency', D.Phil. thesis for presentation to the University of York in 1981 and publication in London in 1982.

Brookshaw, George, *Groups of Flowers drawn and accurately coloured after Nature, with full instructions for the Young Artist,* London, 2nd edn, 1819.

Brookshaw, George, *Groups of Fruit,* London, 2nd edn, 1819.

Brown, Edward, *A brief Account of some Travels in divers parts of Europe,* London, 2nd edn, 1685.

Browne, Alexander, *The whole Art of Drawing, Painting, Limning and Etching,* London, 1660.

Browne, Alexander, *Ars Pictoria: or an Academy treating of Drawing, Painting, Limning, Etching,* London, 1669; 2nd edn, 1675.

Bryant, Joshua, *Bryant's Treatise on the Use of Indian Ink and Colours,* London, n.d. [1808]. [Library of Congress, U.S.A.; microfilm in the possession of the author.]

Cennino d'Andrea Cennini, *Il libro dell'arte,* translated D. V. Thompson as *The Craftsman's Handbook,* New Haven, 1933.

Chambers, Ephraim, *Cyclopaedia: or, an Universal Dictionary of Arts and Sciences,* 2 vols, 1728.

Chaptal, Jean Antoine Claude, *Chimie appliquée aux arts,* 5 vols, Paris, 1807.

Church, Arthur Herbert, *The Chemistry of Paints and Painting,* London, 3rd edn, 1901.

Clark, John Heaviside, *A Practical Essay on the Art of Colouring and Painting Landscapes in Water Colours,* London, 1807.

Clow, Arthur and Nan Louise, *The Chemical Revolution: a Contribution to Social Technology,* London, 1952.

Cobalt Monograph, Centre d'Information du Cobalt, Brussels, 1960.

Colour Index, Society of Dyers and Colourists, Bradford, Yorks., and American Association of Textile Chemists and Colorists, Lowell, Mass., 4 vols, 2nd edn, 1956.

A Compendium of Colors, and other Materials used in the Arts dependent on Design, London, n.d. The book is vol. ii of *The Artist's Repository,* 7th edn, 1797.

Constant-Viguier, *Nouveau manuel complet de miniature, de gouache, du lavis à la sépia, de l' aquarelle et de la peinture à la cire,* Paris, 1845.

Coremans, Paul, Gettens, R. J., and Thissen, Jean, 'La technique des primitifs flamands', *Studies in Conservation,* London, i, 1952, 1–29.

Courtauld Institute of Art, *Exhibition of Books on the Practice of Drawing and Painting from 1650–1850 . . . Dec. 15th–Jan. 11th 1934/5.* [Stencilled catalogue.]

Cox, David, *A Treatise on Landscape Painting in Water Colours,* London, 1814.

Crosland, Maurice P., *Historical Studies in the Language of Chemistry,* London, 1962.

M.D.*** [P.F. Dietrich, Baron d'Holbach], *L'art de la verrerie de Neri, Merret et Kunckel,* Paris, 1752.

Dagley, Richard, *A Compendium of the Theory and Practice of Drawing and Painting,* London, 1818, 2nd edn, 1822.

Dana, James Dwight, *The System of Mineralogy,* 3 vols, New York, and London, 7th edn, 1944–51.

Dayes, Edward, *The Works of the late Edward Dayes,* London, 1805.

Deer, W. A., Howie, R. A., and Zussman, J., *Rock-forming Minerals,* 5 vols, London, 1962–63.

Delaval, Edward Hussey, *An experimental Inquiry into the Cause of the Changes of Colours in opake and coloured Bodies,* London, 1777.

Dictionarium Polygraphicum: or the whole body of arts, regularly digested, ed. John Barrow, 2 vols, London, 1735.

Diderot, Denis, and d'Alembert, Jean le Rond, *Encylopédie: ou dictionnaire raisonné des sciences, des arts et des métiers,* 17 vols, Paris and Neuchâtel, 1754–65.

Dodd, George, *Days at the Factories: or the Manufacturing Industry of Great Britain described,* London, 1843.

Dodd, George, *Textile Manufactures,* 6 vols, London, 1844–46.

Dossie, Robert, *The Handmaid to the Arts,* 2 vols, London, 1758; 2nd edn, 1764.

Dossie Robert, *Memoirs of Agriculture and other Oeconomical Arts,* 3 vols, London, 1768–82.

Dufour, Augustine, *L'art de peindre les fleurs à l'aquarelle,* Paris, 1834.

East India Company, *Letters received by the East India Company from its servants in the East,* ed. William Foster, London, 1896–1902.

East India Company, *A Calendar of the Court Minutes, etc. of the East India Company,* ed. E. B. Sainsbury, Oxford, 1907.

Eastlake, Charles Lock, *Materials for a History of Oil Painting,* London, 1847; 2nd edn, 2 vols, 1869. Reprinted as *Methods and Materials of the Great Schools and Masters,* New York, 1960.

Edwards, W. H., *The Young Artist's Guide to Flower Drawing, and Painting in Water Colours,* London, 1820.

Elsum, John, *The Art of Painting after the Italian Manner,* London, 1704.

The Excellency of the Pen and Pencil exemplifying the uses of them in the most exquisite and mysterious Arts of Drawing, Etching, Engraving, Limning, Painting in Oyl, Washing of Maps and Pictures, London, 1688.

Feller, Robert L., 'Studies on the Darkening of Vermilion by Light', *National Gallery of Art Report and Studies in the History of Art 1967,* Washington, 1968, 99–111.

Field, George, *Chromatics: or, an Essay on the Analogy and Harmony of Colours,* privately printed, 1817.

Field, George, *Chromatography: or a Treatise on Colours and Pigments, and of their Powers in Painting, &c.,* London, 1835; also ed. Thomas W. Salter, London, n.d. [1869], and ed. J. Scott Taylor, London, 1885.

Fielding, Theodore Henry, *Index of Colours and Mixed Tints,* London, 1830.

Fielding, Theodore Henry, *On the Theory of Painting, to which is added an Index of Mixed Tints,* London, 2nd edn, 1836; 3rd edn, 1842.

Fielding, Theodore Henry, *Ackermann's Manual of Colours, used in the different branches of Water-Colour painting: with an ample description of the value and properties of each colour,* London, 1844.

Gartside, M., *An Essay on Light and Shade, on Colours, and on Composition in general,* London, 1805.

Gerard, John, *The Herball, or generall historie of plantes,* London, 1597; another edition, 1633.

Gerard, John, *A Catalogue of plants cultivated in the garden of J. Gerard, in the years 1596–1599,* ed. B. D. Jackson, privately printed, 1876.

Gettens, Rutherford, J., Feller, R. L., and Chase, W. T., 'Vermilion and cinnabar', *Studies in Conservation,* London, xvii, 1972, 45–69.

Gettens, Rutherford J., and FitzHugh, E. W., 'Azurite and blue verditer', *Studies in Conservation,* London, xi, 1966, 54–61.

Gettens, Rutherford J., and FitzHugh, E. W., 'Malachite and green verditer', *Studies in Conservation,* London, xix, 1974, 2–23.

Gettens, Rutherford J., Kühn, H., and Chase, W. T., 'Lead white', *Studies in Conservation,* London, xii, 1967, 125–39.

Gettens, Rutherford J., and Stout, G. L., *Painting Materials: a Short Encylopaedia,* New York, 1942; reprinted 1943, 1946 and 1966.

Goeree, Willem, *Verligterie-Kunde, of regt Gebruik der Water Verwen,* Amsterdam, 1670; reprinted 1697. The book is a revised edition of Geerard ter Brugge, *Verligtery Kunst-Boeck,* Leyden, 1634.

Georee, Willem, *Illuminir-oder Erleuchterey-Kunst, oder der rechte Gebrauch der Wasserfarben,* Hamburg, 1677. German translation of *Verligterie-Kunde.*

Goeree, Willem, *The Art of Limning in which the true grounds and perfect use of*

Water-Colours with all their Proprieties, are clearly and perfectly taught, London, 1674. English translation of *Verligterie-Kunde.*

Goeree, Willem, *Inleyding tot de Praktyk der algemeene Schilderkonst,* Amsterdam, 1697. Generally bound with *Verligterie-Kunde* but with separate pagination.

Goeree, Willem, *Answeisung zu der Practic oder Handlung der allgemeinen Mahler-Kunst,* Hamburg, 1678. German translation of the preceding.

Goeree, Willem, *An Introduction to the General Art of Drawing,* London, 1674. English translation.

Graaf, Johannes Alexander van de, *Het de Mayerne Manuscript als Bron voor de Schildertechniek van de Barok,* Mijdrecht, 1958.

Graaf, Johannes Alexander van de, 'The Interpretation of old Painting Recipes', *Burlington Magazine,* London, civ, 1962, 471–75.

Hamilton, George, *The Elements of Drawing in its various branches,* London, 1812.

Hanson, N. W., 'Some painting materials of J. M. W. Turner', *Studies in Conservation,* London, i, 1954, 162–73.

Harley, Rosamond D., 'Literature on technical aspects of the Arts. Manuscripts in the British Museum', *Studies in Conservation,* London, xiv, 1969, 1–8.

Harley, Rosamond D., 'Oil colour containers: development work by artists and colourmen in the nineteenth century', *Annals of Science,* xxvii, 1971, 1–12.

Harley, Rosamond D., 'Background and development of the artist colourmen', *Colour Review,* Winsor & Newton Ltd, Harrow, 1973, 3–5.

Harley, Rosamond D., 'Artists' brushes—historical evidence from the sixteenth to the nineteenth century', *Conservation and Restoration of Pictorial Art* (eds. N. Brommelle and P. Smith), London, 1976, 61–66.

Harley, Rosamond D., 'A nineteenth-century book of colour samples', *Bulletin,* American Institute for Conservation of Historic and Artistic Works, Washington, D.C., xv, 1975, 49–53.

Harley, Rosamond D., 'Field's manuscripts: early nineteenth-century colour samples and fading tests', *Studies in Conservation,* London, xxiv, 1979, 75–84.

Harris, Moses, *Natural System of Colours,* London, 1811.

Harrison, A. W. C., *The Manufacture of Lakes and precipitated Pigments,* eds. J. S. Remington and W. F. Francis, London, 2nd edn, 1957.

Haydocke, Richard, *A Tracte containing the Artes of curious Paintinge, Caruinge & Buildinge* (translation of G. P. Lomazzo, *Trattato dell'arte de la pittura*), Oxford, 1598.

Heaton, Noel, *Outlines of Paint Technology,* London, 3rd edn, 1947.

Henderson, Peter Charles, *The Seasons, or flower-garden: . . . with a treatise, or general instructions for drawing and painting flowers,* London, 1806.

Hilliard, Nicholas, *A Treatise concerning the Arte of Limning together with A More Compendious Discourse concerning ye Art of Liming by Edward Norgate,* eds. R. K. R. Thornton and T. G. S. Cain, Mid Northumberland Arts Group, Ashington, and Carcanet New Press, Manchester, 1981.

Hodson, T., and Dougall, I., *The Cabinet of the Arts: being a new and universal Drawing Book,* London, 1805.

Hoofnail, John, *New Practical Improvements, and Observations on some of the Experiments . . . of Robert Boyle, Esq. so far as they relate to Tinctures and Pigments,* London, 1738; 2nd edn, [1742?]; later reprinted as *The Painter's Companion,* London, 5th edn, 1815.

Houghton, John, *A Collection for the Improvement of Husbandry and Trade,* London, 1727–28.

Hudson, D., and Luckhurst, K. W., *The Royal Society of Arts,* London, 1954.

Hurry, J. B., *The Woad Plant and its Dye,* London, 1930.

Hurst, George, H., *Painters' Colours, Oils and Varnishes,* London, 1892.

Ibbetson, Julius Caesar, *Process of Tinted Drawing,* privately printed, n.d. [1794].

Ibbetson, Julius Caesar, *An Accidence or Gamut of Painting in Oil and Water Colours,* London, 1803.

Kearney, Hugh (ed.), *Origins of the Scientific Revolution,* London, 1964.

Kühn, Hermann, 'Une boîte à couleurs française du débout du XIXe siècle', *Peintures, Pigments, Vernis,* Paris, xxxix, 1963, 464–77.

Kühn, Hermann, 'Lead-tin yellow', *Studies in Conservation,* London, xiii, 1968, 7–33.

Kühn, Hermann, 'Verdigris and copper resinate', *Studies in Conservation,* London, xv, 1970, 12–36.

Kunckel, Johann, *Ars vitraria experimentalis, oder vollkommene Glasmacher-Kunst,* Frankfurt and Leipzig, 1679; 2nd edn, 1689; 3rd edn, Nuremberg, 1743 and 1756. See also, M. D., Merret and Neri.

Laurie, Arthur Pillans, *The Pigments and Mediums of the Old Masters,* London, 1914.

Leggett, William F., *Ancient and Medieval Dyes,* Brooklyn, 1944.

Lewis, William, *Commercium philosophico-technicum: or, the Philosophical Commerce of Arts: designed as an attempt to improve Arts, Trades and Manufactures,* London, 1763.

[*Limming*], *A very proper treatise, wherein is briefly sett forthe the arte of Limming,* London, 1573; reprinted 1581, 1583, 1588, 1596, 1605 and Michigan, 1932. See also *Technical Studies in the Field of the Fine Arts,* ii, 1933, 35–37, for a review of the 1932 reprint and a discussion of related manuscript material by D. V. Thompson.

Lomazzo, Giovanni Paolo, *Trattato dell'arte de la pittura,* Milan, 1584. See Haydocke for English translation.

Mackenzie, Colin, *One Thousand Processes in Manufactures and Experiments in Chemistry,* London, 5th edn, 1825.

MacTaggart, Peter and Ann, 'Refiners' verditer', *Studies in Conservation,* London, xxv, 1980, 37–45.

Le maître de miniature, de gouache et d'aquarelle, Paris, 1820.

Massoul, Constant de, *A Treatise on the Art of Painting and the Composition of Colours,* London, 1797.

Mérimée, M. J. F. L., *De la peinture à l'huile,* Paris, 1830.

Mérimée, M. J. F. L., *The Art of Painting in Oil, and in Fresco,* London, 1839. English translation of the preceding.

Merret, Christopher, *The Art of Glass, wherein are shown the wayes to make and colour Glass, Pastes, Enamels, Lakes, and other Curiosities,* London, 1662. See also M. D., Kunckel and Neri.

Merrifield, Mary P., *Original Treatises dating from the XIIth to XVIIIth centuries on the Arts of Painting,* 2 vols, London, 1849; reprinted, New York, 1967.

Mühlethaler, Bruno, and Thissen, J., 'Smalt', *Studies in Conservation,* London, xiv, 1969, 47–61.

Neri, Antonio, *L'arte vetraria,* Florence, 1612. Translated into English by Merret, 1662, German by Kunckel, 1679, and French by Dietrich, 1752, all of whom amplified preceding editions.

Nicholson, Francis, *The Practice of Drawing and Painting Landscape from Nature in Water Colours,* London, 2nd edn, 1823.

Norgate, Edward, *Miniatura or the Art of Limning,* ed. Martin Hardie, Oxford 1919.

Norgate, Edward, *A More Compendious Discourse concerning ye Art of Liming,* see Hilliard, eds. R. K. R. Thornton and T. G. S. Cain, Ashington and Manchester, 1981.

Pacheco, Francisco, *Arte de la Pintura. Edición del manuscrito original acabado el 24 de enero de 1638,* ed. F. J. Sanchez Canton, 2 vols, Madrid, 1956.

The Painter's Companion, or, a Treatise on Colours, London, 5th edn, 1815. See Hoofnail.

Parkes, Samuel, *Chemical Essays, principally relating to the Arts and Manufactures of the British dominions,* 5 vols, London, 1815.

Parkinson, John, *Theatrum Botanicum, the Theater of Plantes or an universal and compleate Herball,* London, 1640.

Partington, James Riddick, *A History of Chemistry,* London, vols ii–iv, 1961–64, in progress.

Payne, John, *The Art of Painting in Miniature, on Ivory,* London, 2nd edn, 1798; 6th edn, 1812.

Peacham, Henry, *The Art of Drawing with the Pen, and Limming in Water Colours, more exactlie then heretofore taught and enlarged: with the true manner of painting upon glasse, . . . making your furnace, annealinge, &c.,* London, 1606.

Peacham, Henry, *The Gentlemans Exercise. Or an exquisite practise, as well for drawing of all manner of Beasts in their true Portraittures: as also the making of all kinds of colours, to be vsed in Lymming, Painting, Tricking, and Blason of Coates, and Armes,* London, 1612.

Peacham, Henry, *Graphice or the most ancient and excellent Art of Drawing and Limming,* London, 1612.

Peacham, Henry, *The Compleat Gentleman,* London, 1622; 2nd edn, 1634; 3rd edn, 1661. Reprinted with an introduction by G. S. Gordon, Oxford, 1906.

Perkin, A. G., and Everest, A. E., *The Natural Organic Colouring Matters,* London, 1918.

Pernety, Antoine-Joseph, *Dictionnaire portatif de peinture, sculpture et gravure,* Paris, 1757.

Plesters, Joyce, 'Ultramarine Blue, Natural and Artificial', *Studies in Conservation,* London, xi, 1966, 62–81.

Plesters, Joyce, 'A preliminary note on the incidence of discolouration of smalt in oil media', *Studies in Conservation,* London, xiv, 1969, 62–74.

Pomet, Pierre, *Histoire générale des drogues, traitant des plantes, des animaux et des mineraux etc.,* Paris, 1694.

Pomet, Pierre, *A compleat history of Drugs, written in French by . . . Pomet . . . to which is added what is further observable on the same subject from . . . Lemery and Tournefort, divided into three classes, vegetable, animal and mineral,* 2 vols., London, 1712.

Practical Treatise on Painting in Oil-Colours, London, 1795.

Pretty, Edward, *A Practical Essay on Flower Painting in Water Colours,* London, 1810. [Royal Horticultural Society, London.]

Pulsifier, William H., *Notes for a History of Lead and an inquiry into the development of the manufacture of white lead and lead oxides,* New York, 1888.

Purchas, Samuel, *Pvrchas his Pilgrimage,* London, 1614.

Purchas, Samuel, *Pvrchas his Pilgrimes,* 5 vols, London, 1625.

Rabaté, Henri, 'Sur l'orde chronoligique d'apparition des principaux constituants solides des peintures et des préparations assimilées. 1. Pigments blancs et matières de charge blanches', *Peintures, Pigments, Vernis,* Paris, xl, 1964, 525–27, 601–04, 667–69, 731–32 and xli, 1965, 1–2.

Ray, John, *Observations, topographical, moral & physiological; made in a journey through part of the Low-countries, Germany, Italy, and France: with a catalogue of plants not native in England, found spontaneously growing in those parts, and their virtues,* London, 1673.

Ray, John, *A Collection of English Words not generally used, . . . and an account of the preparing and refining such metals and minerals as are gotten in England,* London, 1674.

Remington, J. S., and Francis, W., *Pigments: their manufacture and use,* London, 3rd edn, 1954.

Roberts, James, *Introductory Lessons, with familiar examples in Landscape, for use of those who are desirous of gaining some knowledge of the pleasing art of painting in water colours,* privately printed, 1809.

Rose, Friedrich, *Die Mineralfarben und die durch Mineralstoffe erzeugten Färbungen,* Leipzig, 1916.

Royal Society of London, *Catalogue of Scientific Papers,* 19 vols, London, 1867–1925.

Salmon, William, *Polygraphice, or the Art of Drawing, Engraving, Etching, Limning, Painting, Washing, Varnishing, Colouring and Dying,* London, 1672; 2nd edn, 1673; 3rd edn, 1675; 5th edn, 1681, 1685; 8th edn, 2 vols, 1701.

Sanderson, John, *The Travels of John Sanderson in the Levant 1584–1602,* ed. William Foster, Hakluyt Society, London, 1931.

Sanderson, William, *Graphice: the use of the Pen and Pencil or, the most excellent Art of Painting,* London, 1658.

Scheele, Carl Wilhelm, *Collected Papers,* translated L. Dobbin, London, 1931.

Scheffer, Johann, *Graphice: id est de arte pingendi,* Nuremberg, 1669.

A Series of progressive Lessons intended to elucidate the Art of Painting in Water Colours, London, 1811; 2nd edn, 1812.

Shaw, Peter, *Chemical Lectures, publickly read at London, in the years 1731, and 1732; and since at Scarborough, 1733; for the Improvement of Arts, Trades, and Natural Philosophy,* London, 1733; reprinted 1755.

Shellac, Angelo Brothers Limited, Calcutta, 1965.

Singer, Charles, Holmyard, E. J., and Hall, A. R. (eds.), *A History of Technology,* 5 vols, Oxford, 1954–58.

Smith, *A short and direct Method of Painting in Water-Colours,* privately printed, 1730.

Smith, John, *The Art of Painting: wherein is included the whole art of vulgar painting, according to the best and most approved rules for preparing and laying on of Oyl Colours,* London, 1676.

Smith, John, *The Art of Painting in Oyl,* London, 2nd edn, 1687.

Smith, John, *The Art of Painting in Oyl . . . to which is added, the whole art and mystery of colouring maps and other prints with water colours,* London, 3rd edn, 1701; 4th edn, 1705; 5th edn, 1738; 6th edn, 1753.

Smith, John, *Smith's Art of House-Painting,* London, 1821.

Smith, Marshall, *The Art of Painting according to the Theory and Practise of the best Italian, French, and Germane Masters,* London, 1692. [Harvard University Library. Microfilm in the possession of the author.]

Sowerby, James, *A new elucidation of colours, original prismatic, and material; showing their concordance in three primitives, yellow, red and blue; and the means of producing, measuring and mixing them: with some observations on the accuracy of Sir Isaac Newton,* London, 1809.

Talley, Mansfield Kirby, *Portrait painting in England: studies in the technical literature before 1700*, Paul Mellon Centre for Studies in British Art, 1981.

Talley, Mansfield Kirby, and Groen, Karin, 'Thomas Bardwell and his Practice of Painting: a comparative investigation between described and actual painting technique', *Studies in Conservation,* London, xx, 1975, 44–108.

Taylor, F. Sherwood, *A History of Industrial Chemistry,* London, 1957.

Taylor, J. Scott, *A descriptive Handbook of modern Water-Colour Pigments,* London, n.d. [1887].

Thompson, Benjamin, Count Rumford, *Proposals for forming by subscription, in the Metropolis of the British Empire, a public Institution for diffusing the knowledge and facilitating the general introduction of useful mechanical inventions and improvements, and for teaching, by courses of philosophical lectures and experiments, the application of science to the common purposes of life,* privately printed, 1799.

Thompson, Daniel Varney, and Hamilton, George Heard, *An anonymous fourteenth-century treatise: De arte illuminandi,* New Haven and London, 1933.

Thompson, Daniel Varney, *The Materials of Medieval Painting,* London, 1936. republished 1956.

Thompson, Daniel Varney. See also Cennino and *Limming.*

Tingry, Pierre François, *The Painter and Varnisher's Guide,* London, 1804.

Towne, Thomas, *The Art of Painting on Velvet,* privately printed, 1811.

Ure, Andrew, *A Dictionary of Arts, Manufactures, and Mines,* London, 1839; 3rd edn, 1843.

Varley, John, *J. Varley's List of Colours,* London, 1816.
Waller, Richard, 'A catalogue of simple and mixt colours, with a specimen of each colour prefixt to its proper name', *Philosophical Transactions,* London, xvi, 1686, 24–32.
Watin, *L'art du peintre, doreur, vernisseur,* Paris, 1772; 4th edn, 1787.
Watson, Richard, *Chemical Essays,* 5 vols, Cambridge, 1781–88.
Wild, A. Martin de, *The Scientific Examination of Pictures,* London, 1929.
Williams, W., *An Essay on the Mechanic of Oil Colours,* Bath, 1787.
Winsor, William, and Newton, Henry Charles, *Remarks on the White Pigments used by Water-Colour Painters, being a letter addressed to G. H. Bachhoffner, Esq. . . . relative to a passage in his treatise on Chemistry as applied to the Fine Arts,* privately printed, 1837.
Winsor, William, *The Hand-Book of Water-Colours, a brief treatise on their qualities and effects when employed in painting, with some account of the general nature of colours,* privately printed, n.d. [*c.* 1846/57].
The Young Artist's Complete Magazine, London, n.d. [*c.* 1785?].

References

Unless otherwise stated, works in English were published in London and those in French were published in Paris.

1. The manuscripts are discussed in R. D. Harley, 'Literature on Technical Aspects of the Arts. Manuscripts in the British Museum', *Conservation,* xiv (1969), 1–8. It includes folio references for the parts on painting and gives an indication of other subjects covered by the manuscripts.
2. Bodleian MS. Ashmole 1491 includes information on art subjects: p. 1208 (vermilion), p. 1210 (varnish) and p. 1289 (verdigris). Bodleian MS. Ashmole 1494, p. 116 (azure) and p. 636 (list of colours and gums).
3. The book was reprinted many times, in 1581, 1583, 1588, 1595, 1605, 1615, and in the Michigan Facsimile Series, No. 3 (Ann Arbor Press, Michigan), 1932.
4. Nostell MS. Vertue (bundle II), f. 120. Quoted from *Walpole Soc.* (Oxford), xxx (1952), 49.
5. Edinburgh University MS. Laing III 174, f. 1. An exact transcript together with a modern version is provided in Nicholas Hilliard, *A Treatise concerning the Arte of Limning,* eds. R. K. R. Thornton and T. G. S. Cain (Mid Northumberland Arts Group, Ashington), 1981.
6. H. Peacham, *Peacham's Compleat Gentleman 1634* (Oxford, 1906). The introduction by G. S. Gordon contains details of Peacham's career.
7. H. Peacham, *The Compleat Gentleman* (1634), 65.
8. H. Peacham, *The Compleat Gentleman* (1634), 129–30.
9. T. Gibson, 'A Sketch of the Career of Theodore Turquet de Mayerne', *Annals of Medical History* (New York), new series v (1933), 315–26. See also 'The Iconography of Sir Theodore de Mayerne', *Annals of Medical History* (New York), 3rd series iii (1941), 288–96.
10. E. Berger, *Beiträge zur Entwicklungs geschichte der Maltechnik* (Munich), iii (1912).
    J. A. van de Graaf, *Het de Mayerne Manuscript als Bron voor de Schildertechniek van de Barok* (Mijdrecht, 1958).
    A partial transcript of B.M. MS. Sloane 2052 is also printed in *P.P.V.,* xli (1965), and xlii (1966).
11. F. W. Gibbs, 'An Account of a Manuscript entitled "Saponis Artificium" ', *Chemistry and Industry,* lvii (1938), 877. The author suggests that B.M. MS. Sloane 1990 (which contains details of soap-making on f. 127) is that which was brought to the attention of the Royal Society in the seventeenth century when the Fellows were investigating various trades and manufactures. It seems that it was then in the possession of Sir Theodore de Vaux and that it, and possibly B.M. MS. Sloane 2052, may have passed into the hands of Sir Hans Sloane at that time, as he acquired many of the trade papers while President. The subsequent neglect and rediscovery of the manuscript in

recent times is discussed by A. E. Werner, 'A "New" de Mayerne Manuscript', *Conservation,* ix (1964), 130–33.

12. E. Norgate, *Miniatura or the Art of Limning,* ed. M. Hardie (Oxford, 1919), 5.
13. E. Norgate, *Miniatura or the Art of Limning,* ed. M. Hardie (Oxford, 1919), v–xxix. The original treatise was written after 1621 (when the Earl of Arundel became Earl Marshal) and during or before 1626 (the death of Paolo Bril). The dates of the revised treatise can be set between December 1648 (the death of Peter Oliver) and October 1650 (the death of Francis Cleyn, the younger).
14. B.M. MS. Additional 23073, f. 44. Quoted from *Walpole Soc.,* xxvi (1938), 60.
15. For Gyles' career, see J. A. Knowles, 'Henry Gyles, Glass-Painter of York', *Walpole Soc.,* xi (1923), 47–72. B.M. MS. Harley 6376, p. 111, contains a reference to Gyles Vermulen, a Dutchman, in a recipe dated 1679; the same man is mentioned in two letters to Henry Gyles from a friend in London, making it clear that Gyles had written to Vermulen several times to ask for information (B.M. MS. Stowe 746, f. 57 and f. 59). Vertue's comments occur in B.M. MS. Additional 23070, f. 58, quoted from *Walpole Soc.,* xx (1932), 66. A particularly interesting feature of Gyles' book (B.M. MS. Harley 6376) is the acount of artists' brush manufacture which is original and does not appear in any of the other books that owe so much to Norgate's treatise. See R. D. Harley, 'Artists' brushes—historical evidence from the sixteenth to the nineteenth century', *Conservation and Restoration of Pictorial Art,* eds. N. Brommelle and P. Smith (1976), 61–66.
16. Mary R. S. Beal, 'A study of Richard Symonds: his Italian notebooks and their relevance to seventeenth-century painting techniques', unpublished Ph.D. thesis, University of London, 1978. Contains a transcript of B.M. MS. Egerton 1636.
17. F. Hard, 'Richard Haydocke and Alexander Browne: Two Half-forgotten Writers on the Art of Painting', *P.M.L.A.* (New York), lv (1940), 727–41.
18. A. Browne, *Ars Pictoria* (1675), 39.
19. Patent Office, printed series No. 421.
20. Several attributions are discussed in Constant-Viguier *et al., Nouveau manuel complet de miniature* (1845), 24. There it is stated that the attribution to Claude Boutet is based on a manuscript note in a copy belonging to the Bibliothèque du Roi (now the Bibliothèque Nationale) and that the attribution to Ballard is preferable.
21. The name Hoofnail is very unusual and suggests an association with the Hoefnagel family, of Flemish origin, known as painters and engravers in the sixteenth and seventeenth centuries. It is interesting but possibly coincidental to find that a John Hoefnagel was naturalised in London in 1710. *Publications of the Huguenot Society* (Manchester), xxvii (1923), 105.
22. M. Kirby Talley and Karin Groen, 'Thomas Bardwell and his *Practice of Painting*: a comparative investigation between described and actual painting technique', *Conservation,* xx (1975), 44–108, and R. White, 'An examination of Thomas Bardwell's portraits — the media', *Conservation,* (1975), 109–13. For repetition of Bardwell's writing, see F. Schmid, 'The strange case of Thomas Bardwell', *Technical Studies in the Field of the Fine Arts* (Harvard), ix (1941), 153–59.
23. W. Williams, *An Essay on the Mechanic of Oil Colours* (Bath, 1787), 51.
24. The manuscript notebooks passed from Field to Henry Charles Newton, one of the partners in Winsor & Newton, the British artists' colour firm established in 1832. They remained with the Newton family until 1973 when they were placed in the keeping of the Courtauld Institute of Art, University of London. These manuscripts, which form a very important primary source for the history of pigments in the early nineteenth century, were not available for study when the first edition of *Artists' Pigments* c. *1600–1835* was in preparation, hence the reliance on *Chromatography* as a source for the first edition. It is now clear that, in *Chromatography,* Field summarised the

information he had gathered and judgements he had formed while undertaking the experiments that are reported in the notebooks, and, for this reason, many references to *Chromatography* have been retained for the present edition. Where useful additional information is contained in the manuscripts this has been included, as also are references to colour samples in one particular notebook when they are of particular interest. This manuscript (Courtauld Institute, Field/6) is referred to in the text by the abbreviated title 'Practical Journal 1809'.

For further discussion of the manuscripts see R. D. Harley, 'Field's manuscripts: early nineteenth-century colour samples and fading tests', *Conservation,* xxiv (1979), 75–84, and 'A nineteenth-century book of colour samples', *Bulletin,* American Institute for Conservation of Historic and Artistic Works (Washington, D.C.), xv (1975), 49–53. The former provides a list of the artists with whom Field was connected. For external evidence that Field supplied both both Lawrence and Millais, see C. L. Eastlake, *Methods and Materials of Painting of the Great Schools and Masters* (New York, 1960; reprint of the first edition 1847), i, 446, and W. Holman Hunt, *Pre-Raphaelitism* (1913), 306.

25. A list of the volumes with the Public Record Office class references is supplied in Appendix One with the list of other Public Record Office references and the abbreviations used throughout. That list provides sufficient information for any reader who wishes to follow up comments concerning imports which are made in subsequent chapters; consequently, detailed references are not given in the text. The first series of ledgers up to and including 1780 contains details of imports and exports, London being listed separately from the outports. The search was restricted to entries for London, because the volume of goods entering the capital far exceeded that for any other port. In the first series, imports are listed under the country of origin with details of quantity and value which are sub-divided according to the nationality of the ships, either British or foreign. The series of ledgers which includes those for 1792 and 1800 has imports only, the commodities being listed in alphabetical order with the countries of origin and quantities listed under each heading. The remaining ledgers, 1810–30, are included in a different series containing imports only, but various commodities are once again listed under the country of origin.
26. *Letters received by the East India Company from its servants in the East,* ed. W. Foster (1896–1902).
    A Calendar of the Court Minutes, etc., of the East India Company, ed. E. B. Sainsbury (Oxford, 1907).
27. *Calendar of State Papers and Manuscripts relating to English Affairs, existing in the Archives and Collections of Venice, and in other Libraries of northern Italy,* ix-xx (1592–1626) are particularly useful for the declining importance of trade in the Mediterranean.
28. T. Fairman Ordish, 'Early English Inventions', *Antiquary,* xii (1885), 1–6, 61–65, 113–18.
    E. Wyndham Hulme, *The Early History of the English Patent System* (Boston, 1909), 122–38.
    R. Jenkins, 'The Protection of Inventions during the Commonwealth and Protectorate', *Notes and Queries,* 11th series vii (1913), 162–63.
    A. A. Gomme, 'Date Corrections of English Patents, 1617–1752', *Transactions of the Newcomen Society,* xiii (1932–33), 159–64.
    Apart from the omission of certain patents from the printed series, a further complication arises from the fact that, when the series was compiled, the different conventions in regard to the beginning of a new year (25th March before 1752 and 1st January thereafter) were ignored, so that by modern standards many of the printed sources are wrongly dated. For this work, therefore, the manuscript sources at the Public Record Office were used for

all patents earlier than 1752. References are given to the patent rolls and also the Patent Office printed series number where applicable. The omission of any mention of the printed series in a reference may be taken as an indication that the patent concerned is not included in the series of blue books. In accordance with modern practice, the day and month in any date is retained even though it may be old style, but the year is given in modern form. All the patents mentioned were valid in England and Wales; research was not extended to patents which had been enrolled in Scotland.

29. F. Bacon, *The Advancement of Learning* (1605), quoted from *Origins of the Scientific Revolution,* ed. H. Kearney (1964), 121.
30. Royal Society, Classified Papers iii (1), 1.
31. T. Birch, *The History of the Royal Society* (1756 and 1757), ii, 231.
32. Two of Hooke's contemporaries describe how he had shown promise in drawing and painting as a boy and had been encouraged by the younger Hoskins, after which he was apprenticed to Sir Peter Lely. According to John Aubrey, Hooke was sent up to London with £100 'with an intention to have bound him apprentice to Mr. Lilly, the paynter, with whom he was a little while upon tryall, who liked him very well, but Mr. Hooke quickly perceived what was to be donne, so, thought he, why cannot I doe this by myselfe and keepe my hundred pounds? He also had some instruction in drawing from Mr. Samuel Cowper (prince of limners of this age).' Quoted from R. T. Gunther, *Early Science in Oxford,* vi (privately printed, 1930), 5. A similar account was given by Richard Waller, who also mentioned that the smell of oil colours disagreed with Hooke. R. Waller, *Posthumous Works of Robert Hooke* (1705), iii.
33. M. 'Espinasse, *Robert Hooke* (1956), 27.
34. *Phil. Trans.,* xvi (1686), 24–28.
35. F. W. Gibbs, 'Robert Dossie (1717–77) and the Society of Arts', *Annals of Science,* vii (1951), 149–72 and ix (1953), 191–93.
36. *Transactions of the Society of Arts,* xx (1802), v–vi.
37. B. Thompson, Count Rumford, *Proposals for forming . . . a public institution for diffusing the knowledge and facilitating the general introduction of useful mechanical inventions and improvements* (privately printed, 1799), 7–8.
38. G. Dodd, *Days at the Factories* (1843), 10.
    It is worth noting that Dodd's short description of artists' colourmen was out of date on some points, for moist water colours in pans had been introduced in the 1830s. Also, syringe containers for oil colours had been introduced briefly in 1840, followed by universally adopted collapsible tubes in 1841. See R. D. Harley, 'Oil colour containers: development work by artists and colourmen in the nineteenth century', *Annals of Science,* xxvii (1971), 1–12 and Plates I–X.
39. J. Wood, *A personal Narrative of a Journey to the Source of the River Oxus* (1841), 262–66.
40. R. J. Gettens, 'Lapis Lazuli and Ultramarine in Ancient Times', *Alumni,* ixx (1950), 342–57.
    J. Beckmann, *A History of Inventions and Discoveries* (1814), ii, 315.
41. B.M. MS. Sloane 2052, f. 68.
    J. Payne, *The Art of Painting in Miniature* (1798), 6.
42. B.M. MS. Sloane 2052, f. 70.
43. B.M. MS. Harley 6000, ff. 14v–15.
44. B.M. MS. Sloane 2052, f. 70.
45. J. C. Ibbetson, *An Accidence or Gamut of Painting* (1803), 17.
46. Courtauld Institute, Field/6, f. 334. Field's permanence tests included exposure to light and exposure of a separate sample to impure air. Natural ultramarine is generally regarded as a comparatively permanent pigment. For discussion of possible occasional deficiency, 'ultramarine sickness', see

J. Plesters, 'Ultramarine Blue, Natural and Artificial', *Conservation,* xi (1966), 62–91.
47. R. Waller, 'A Catalogue of simple and mixt Colours', *Phil. Trans.,* xvi (1686), 26.
48. A discussion about *bice* and its derivation is included in *De arte illuminandi,* eds. D. V. Thompson and G. H. Hamilton (New Haven and London, 1933), 58.
49. Evidence for interrupted supplies of azurite from Hungary rests on remarks of the seventeenth-century Spanish painter Pacheco. He describes how Philip II admired a masterpiece by van Eyck, requested the Flemish painter Coxie to make a copy, and, because he could not find a blue as fine as in the original, sent to Titian in Venice for some. Pacheco states that it was a natural blue, found in Hungary, which was more easily obtained before the Turks were overlords of that region. F. Pacheco, *Arte de la Pintura. Edición del manuscrito original acabado el 24 de enero de 1638* (Madrid, 1956), ii, 61.
50. B.M. MS. Sloane 2052, f. 15v, f. 22, f. 86 and f. 94.
Previous writers who have missed the fact that bice from the Indies meant the Spanish Indies, i.e. Latin America, include Eastlake and van de Graaf. The use of Mexican azurite is substantiated by the identification of the pigment in murals of pre-Columbian people of the American Southwest and on a Mexican painting of the colonial period. Even though both occurrences are mentioned by R. J. Gettens and E. W. FitzHugh, 'Azurite and Blue Verditer', *Conservation,* xi (1966), 54–61, Central America has been entirely overlooked as a source of azurite used in Europe.
51. B.M. MS. Sloane 2052, f. 22.
52. H. Peacham, *Graphice* (1612), 84.
53. B.M. MS. Harley 6000, f. 3v.
54. B.M. MS. Harley 6000, f. 5.
55. B.M. MS. Harley 6000, f. 10.
56. Ure's definition of *bice* is quoted in the *Oxford English Dictionary* but there is no reference to the use of the name in the seventeenth century. D. V. Thompson, *The Materials of Medieval Painting* (1936), 152, discusses the meaning of the name in medieval times and the eighteenth century but likewise ignores the intervening period. Other modern writers also tend to overlook documentary sources of the seventeenth century. R. J. Gettens and E. W. FitzHugh, 'Azurite and Blue Verditer', *Conservation,* xi (1966), 54–61, list bice as a synonym for verditer but fail to include it with obsolete terminology for azurite.
57. Instructions, both practicable and impracticable, for making artificial azures are included in the following manuscripts:
B.M. MS. Sloane 288, f. 128, f. 128v and f. 130.
Bodleian MS. Ashmole 1480, f. 12.
Bodleian MS. Ashmole 1494, p. 116.
B.M. MS. Sloane 1394, f. 137v and f. 138.
B.M. MS. Sloane 6284, f. 113v.
B.M. MS. Sloane 1990, f. 20v.
B.M. MS. Sloane 2052, f. 22v, f. 54, f. 65, ff. 66v–67v and ff. 70–71.
B.M. MS. Additional 12461, f. 54.
58. For a survey of various types of manufactured copper blues, see M. V. Orna, M. J. D. Low and N. S. Baer, 'Synthetic blue pigments: ninth to sixteenth centuries. I. Literature', *Conservation,* xxv (1980), 53–63.
59. B.M. MS. Sloane 2052, ff. 65–67v. See also J. A. van de Graaf, *Het de Mayerne Manuscript* (Mijdrecht, 1958), 171.
60. B.M. MS. Sloane 2052, f. 22v.
The parting process employed by silver refiners consisted of immersing plates of impure copper in nitric acid so that any silver contained in the copper was precipitated and settled in a black, pasty mass at the bottom of the vessel, covered meanwhile by copper nitrate, a greenish liquid. In order to

make verditer, whiting was placed in a tub and copper nitrate was poured on it, a chemical reaction took place and after some time the pigment, basic copper carbonate, was left at the bottom of the tub. The almost colourless liquid which covered it was poured off, boiled and treated so that it could once again be used in the parting process.

61. R. Boyle, *Works* (1744), i, 217. The account of verditer is included in the essay 'Concerning the Unsuccessfulness of Experiments', originally printed in 1661.
62. C. Merret, *The Art of Glass* (1662), 292.
63. P. and A. MacTaggart, 'Refiners' Verditers', *Conservation,* xxv (1980), 37–45.
64. *Cyclopaedia,* ed. E. Chambers (1728), *s.v.* Verditer.
65. R. E. Wilson, *A History of the Sheffield Smelting Company Limited, 1760–1960* (1960), 27–28.
66. J. A. C. Chaptal, *Chimie appliquée aux arts* (1807), iii, 415–19.
67. P. and A. MacTaggart, 'Refiners' Verditers', *Conservation,* xxv (1980), 37–45.
68. G. Agricola, *De re metallica,* eds. H. C. and L. H. Hoover (1912), 214.
69. *Cobalt Monograph* (Brussels, 1960), 1–3.
70. J. R. Partington, *A History of Chemistry,* ii (1961), 59–60.
71. J. Kunckel, *Ars vitraria experimentalis* (Frankfurt and Leipzig, 1689), 46–47. For a contemporary English translation, see Royal Society, Classified Papers xx, 95. At one time the manuscript was attributed to Martin Lister in the Society's manuscript index, but the attribution was later altered to Hooke. A previous writer has pointed out that the paper is a translation from Kunckel which was produced by Hooke at a meeting of the Royal Society in 1685. R. T. Gunther, *Early Science in Oxford* (privately printed, 1930), vii, 681. However, others have ignored the connection, and the erroneous attribution to Lister has been perpetuated in *History of Technology* (Oxford, 1957), iii, 703.
72. M. D. [P. F. Dietrich], *Art de la verrerie* (1752), 595–600.
73. B.M. MS. Sloane 2052, f. 9v.
74. For notable occurrences of smalt in paintings, see B. Mühlethaler and J. Thissen, 'Smalt', *Conservation,* xiv (1969), 47–61.
75. Courtauld Institute, Field/6, ff. 315, 343 and 358.
    For a discussion concerning the relative instability of smalt, see J. Plesters, 'A preliminary note on the incidence of discolouration of smalt in oil media', *Conservation,* xiv (1969), 62–74.
76. L. J. Thenard, *Traité de chimie,* ii (6th edn, 1834), 333–44.
    For the history of research on the properties of cobalt see also:
    P.P.V., xxiii (1947), 229–30.
    M. Deribere and C. More, 'Pigments au cobalt', *Chimie des peintures (Brussels), vi (1943).*
    J. R. Partington, A History of Chemistry, ii (1961), 708–09 and iii (1962), 168, 183, 190, 201.
77. J. A. C. Chaptal, *Chimie appliquée aux arts* (1807), iii, 373.
    M. J. F. L. Mérimée, *The Art of Painting* (1839), 153.
78. L. J. Thenard, 'Sur les couleurs, suivies d'un procédé pour préparer une couleur bleue aussi belle que l'outremer', *Journal des mines,* xv (1803–04), 128–36.
79. *P.P.V.,* xxxix (1963), 470–71.
80. S. Parkes, *Chemical Essays* (1815), i, 93.
81. Courtauld Institute, Field/6, f. 352.
    The entry may be dated soon after 1815, the year Sir Thomas Lawrence, who is mentioned frequently in the notebook, was knighted.
82. *Annales de chimie,* xxxvii (1828) 409–13.
83. *Annales de chimie,* xlvi (1831), 431–34. See also J. Plesters, 'Ultramarine Blue, Natural and Artificial', *Conservation,* xi (1966), 62–91 for an account of the discovery of artificial ultramarine.
84. C. L. Eastlake, *Methods and Materials of the Great Schools and Masters* (New York) 1960, i, 453. The extract from Richard Symond (B.M. MS. Egerton 1636,

f. 30) is quoted in Mary R. S. Beal, 'A study of Richard Symonds: his Italian notebooks and their relevance to seventeenth-century painting techniques.' Unpublished Ph.D. thesis, University of London (1978), 101.

85. Berger MS., Stock list and accounts for 1805, p. 8.
86. E. H. Delaval, *An experimental Inquiry into the Cause of Changes of Colours* (1777), 54.
    Nicholson's Journal, 2nd series, i (1802), 153–54.
87. A sample of blue ochre is included in Field's notebook, Courtauld Institute, Field/6, f. 365. See also G. Field, *Chromatography* (1835), 114.
88. J. Gerard, *The Herball* (1633), 336.
89. B.M. MS. Harley 6000, f. 1v.
90. J. Ray, *Observations . . . made in a journey* (1673), 296.
91. R. Boyle, *Experiments and Considerations touching Colours* (1664), 213–16.
92. H. Peacham, *The Gentlemans Exercise* (1612), 85.
93. W. Goeree, *The Art of Limning* (1674), 3.
94. The Statutes banning logwood (23 Elizabeth I, c. 9 and 39 Elizabeth I, c. 11) are correctly interpreted by J. B. Hurry, *The Woad Plant and its Dye* (1930). They are misinterpreted by A. G. Perkin and A. E. Everest, *The Natural Organic Colouring Matters* (1918), 475, and their erroneous statement that the use of indigo was prohibited is quoted by R. J. Gettens and G. L. Stout, *Painting Materials* (New York, 1942), 121.
95. Patents granting the right to import logwood include the following Public Record Office references:
    C.66/1643/32(1604).
    C.66/1677/24(1605).
    C.66/2632/40(1633).
    C.66/2950/41(1660).
    Restrictions were repealed by Statute (13 and 14 Charles II, c. 11).
96. Details of blue lake are omitted from the English translation of Goeree's book but are included in the Dutch and German editions, *Verligterie-Kunde* (Amsterdam, 1697), 16, and *Illuminir-oder Eleuchterey-Kunst* (Hamburg, 1677), 11–12.
97. J. Scheffer, *Graphice* (Nuremberg, 1669), 169.
98. *Letters received by the East India Company*, ed. W. Foster (1896–1902), iii, 179.
99. A difference between the dye obtained from indigo and that from woad has always been recognised, but the latter was not identified until recently. The dye from indigo is indoxyl-ß-D-glucoside, whereas that from woad is indoxyl-5-ketogluconate. See E. Epstein, M. W. Nabors and B. B. Stowe, 'Origin of Indigo of Woad', *Nature,* ccxvi (1967), 547–49.
100. B.M. MS. Sloane 288, f. 130.
101. I.O.R. East India Company, Original Correspondence, iv, 409. Letter from Francis Fettiplace at Agra to the Governor at London, 26th November 1616.
102. S. Purchas, *Purchas his Pilgrimes* (1625), i, 429–30.
103. *The English Factories in India, 1646–50*, ed. W. Foster (Oxford, 1914), 254.
104. B.M. MS. Sloane 2052, f. 4v, f. 94 and f. 144v.
105. B.M. MS. Harley 6000, f. 3.
106. R. J. Gettens and G. L. Stout, *Painting Materials* (New York, 1942), 150.
107. J. C. G. Ackermann, *Das Leben Johann Conrad Dippels* (Leipzig, 1781), 60–68.
108. J. R. Partington, *A History of Chemistry,* ii (1961), 378–9.
109. G. E. Stahl, *Experimenta, Observationes, Animadversiones, CCC Numero, Chymicae et Physicae* (Berlin, 1731), 280–83.
110. *Phil. Trans.,* xxxiii (1724), 15–17. English translation *Phil. Trans.,* abridged series vii (1809), 4–6.
111. P. Shaw, *Chemical Lectures* (1733), 182.
112. G. H. Hurst, *Painters' Colours, Oils, and Varnishes* (1892), 200.
113. I. Bristow, 'Interior house-painting from the Restoration to the Regency', draft of unpublished D. Phil. thesis for the University of York, part 1, chapter 1, 47. The earliest dated formula of Berger's for Prussian blue dates from 1816

in which animal hooves were used rather than blood, and hydrochloric acid was omitted at the end of the process.
114. L. J. M. Coleby, 'A History of Prussian Blue', *Annals of Science,* iv (1939), 206–11.
115. 'Notitia coerulei Berlinensis nuper inventi', *Miscellanea Berlinensia* (Berlin), i (1710), 377–78.
116. *The Repository of Arts,* ii (1809), 3.
117. J. Payne, *The Art of Painting in Miniature* (2nd edn, 1798), 8.
118. Ceder green is discussed by R. D. Harley, 'The Interpretation of Colour Names', *Burlington Magazine,* cx (1968), 460–61.
119. V. Boltz von Rufach, *Illuminirbuch* (1566), 27v.
120. G. Agricola, *De re metallica,* eds. H. C. and L. H. Hoover (1912), 221 and 584.
121. For discussion of malachite, together with a list of notable occurrences in paintings, see R.J. Gettens and E.W. FitzHugh, 'Malachite and green verditer', *Conservation,* xix (1974), 2–23.
122. J. Plesters, 'A technical examination of some panels from Sassetta's Sansepolcro altarpiece', *National Gallery Technical Bulletin,* i (1977), 10–17, and 'A note on the materials, technique and condition of Bellini's "The Blood of the Redeemer"', *National Gallery Technical Bulletin,* ii, (1978), 22–24.
123. For a discussion of nomenclature, variations in composition and notable occurrences in paintings, see H. Kühn, 'Verdigris and copper resinate', *Conservation,* xv (1970), 12–36.
124. B.M. MS. Sloane 1990, f. 51.
J. Ray, *Observations* (1673), 454–56.
125. B.M. MS. Sloane 2052, f. 31 and f. 32.
126. P.R.O. C/3344/7, printed series No. 270 (patent). For the premium see R. Dossie, *Memoirs of Agriculture and other Oeconomical Arts,* i (1768), 176.
127. C. W. Scheele (translated L. Dobbin), *Collected Papers* (1931), 53.
128. C. W. Scheele, *Efterlemnade Bref och Anteckningar,* ed. A. E. Nordenskiöld (Stockholm, 1892), 195.
129. C. W. Scheele (translated L. Dobbin), *Collected Papers* (1931), 195–96.
130. *Journal des mines,* xv (1803–4), 129–30.
131. The discovery of aceto-arsenite of copper is not mentioned in most modern books on artists' colours. Exceptionally, R. Mayer, *The Artist's Handbook* (1964), 55, contains the statement that Scheele himself made the pigment in 1788, but in a private communication Mayer says that he is now unable to supply a reference to substantiate the information.
132. J. von Liebig, 'Sur une couleur verte', *Annales de chimie,* xxiii (1823), 412–13.
133. J. A. C. Chaptal, *Chimie appliquée aux arts* (1807), iii, 420. J. Schroeter, 'History of Inorganic Copper Pigments', *CIBA Review* (Basle), xi, No. 127 (September 1958), 11–14.
134. *Kongl. Vetenkaps Academiens nya Handlingar* (Stockholm), i (1780), 163 and ii (1781), 3. Reference quoted from J. R. Partington, *A History of Chemistry,* iii (1962), 178.
135. A. H. Church, *The Chemistry of Paints and Painting* (1901), 196.
Rinmann's method is also discussed in 'Pigments au cobalt', *Chimie des peintures* (Brussels), vi (1943).
136. *Annales de chimie,* lxx (1809), 70–94.
137. A. H. Church, *The Chemistry of Paints and Painting* (1901), 194.
138. B.M. MS. Stowe 680, f. 132.
139. Anon., *A Book of Drawing* (1666), 13.
140. *P.P.V.,* xxiv (1948), 177–78.
141. M. J. F. L. Mérimée, *The Art of Painting in Oil* (1839), 104.
142. English opinion that Prussian ochre was yellow is emphasised here because van de Graaf suggests that *ochre de Prusse* mentioned by de Mayerne was a burnt, and therefore red, ochre: J. A. van de Graaf, *Het de Mayerne Manuscript* (Mijdrecht, 1958), 23 and 143. His suggestion is based on a remark made by Watin, a French writer of the eighteenth century, whose view that

Prussian ochre was synonymous with red ochre was probably valid for the eighteenth century, for at that time Dutch manufacturers bought ochre from Prussia, converted it to red ochre, and resold it under the name Prussian: *P.P.V.,* xxiv (1948), 177–78. However, it is obviously wrong to explain the terminology of one period by applying a meaning which was current more than a century later, for English sources of the seventeenth century supply incontestable evidence that spruce ochre was a dull yellow. The suggestion that it should be burnt, which is contained in *The Excellency of the Pen and Pencil* (1668), 90, points to the fact that it had not been converted to red ochre before its arrival in England.

143. *P.P.V.,* xxiv (1948), 177–78.
144. B.M. MS. Sloane 1990, f. 26. and f. 29v.
145. A. H. Church, *The Chemistry of Paints and Painting* (1890), 157.
146. Patent Office, printed series Nos. 1243 (1780) and 1996 (1794).
147. G. H. Hurst, *Painters' Colours, Oils and Varnishes* (1892), 142.
148. W. Goeree, *The Art of Limning* (1674), 4.
149. B.M. MS. Sloane 3292, f.3.
150. B.M. MS. Sloane 2052, f. 152 (Johnson) and f. 153v (Van Dyck). R. Haydocke, *Artes of curious Paintinge* (Oxford, 1598), 101. Haydocke translates Lomazzo's suggestion that ground glass should be added to orpiment and adds his own marginal note that it serves as a drier. For the mistaken tradition concerning ground glass, see J. A. van de Graaf, 'The Interpretation of old Painting Recipes', *Burlington Magazine,* civ (1962), 471–75.
151. M. Kirby Talley and K. Groen, 'Thomas Bardwell and his *Practice of Painting:* a comparative investigation between described and actual painting technique', *Conservation,* xx (1975), 44–108.
152. Cennino Cennini (translated D. V. Thompson), *The Craftsman's Handbook* (New York, 1933), 28.
W. Beurs, *De Groote Waereld in 't kleen geschildert etc.* (Amsterdam, 1692), 12v; quoted by van de Graaf, *Het de Mayerne Manuscript* (Mijdrecht, 1958), 51 and 118.
153. M. P. Crosland, *Historical Studies in the Language of Chemistry* (1962), 36.
154. B.M. MS. Sloane 2052, f. 92.
155. M. P. Merrifield, *Original Treatises on the Arts of Painting* (New York, 1967), i, clxii.
156. C. L. Eastlake, *Methods and Materials of the Great Schools and Masters* (New York, 1960), i, 438–39.
157. D. V. Thompson, *The Craftsman's Handbook* (New Haven, 1933), 28. See also D. V. Thompson and G. H. Hamilton, *An anonymous fourteenth-century treatise. De arte illuminandi. The technique of manuscript illumination* (New Haven, 1933), 28.
158. R. Jacobi, 'Über den in der Malerei verwendeten gelben Farbstoff der alten Meister', *Angewandte Chemie* (Berlin), liv (1941), 28–29.
H. Kühn, 'Lead-tin yellow', *Conservation,* xiii (1968), 7–33.
159. Lead monoxide (litharge) has been identified by Lazzarini in some Italian paintings, notably by Sebastiano del Piombo and Cima. See the exhibition catalogue, *Giorgione: La Pala di Castelfranco Veneto,* eds. L. Lazzarini, F. Pedrocco, T. Pignatti, P. Spezzani and F. Valcanover (Castelfranco Veneto), 29 May–30 September 1978, 50. (The author is indebted to Joyce Plesters and Ashok Roy for this reference.)
A list of notable occurrences of lead antimonate (many of them supplied by H. Kühn to J. M. Taylor of the Canadian Conservation Institute) is to appear in I. N. M. Wainwright, J. M. Taylor and R. D. Harley, 'Lead antimonate', part of a study of ten pigments to be published by the National Gallery of Art/National Endowment for the Arts, Washington D.C.
160. Mary R. S. Beal, 'A study of Richard Symonds: his Italian notebooks and their relationship to seventeenth-century painting techniques', unpublished Ph.D. thesis, University of London (1978), 118.

161. J. B. van Helmont, *Oriatrike* (1662), 831. English translation of a treatise in Latin, *De Lithiasi,* written during the first half of the seventeenth century.
162. B.M. MS. Harley 6000, f. 3v.
163. *Repertory of Patents of Invention*, iii (1795), 349–50 (translated from a paper in *Journal de physique*).
164. Fougeroux de Bondaroy, 'Sur le giallolino ou jaune de Naples', *Histoire de l'Académie Royale des Sciences* (1766), 60–64 and 303–14.
165. J. R. Partington, *A History of Chemistry,* iii (1962), 218–19.
166. P.R.O. C.73/16/18, printed series No. 1281. The date is given wrongly as 1780 in *The Repertory of Arts,* xii (1800), 157, which is quoted by J. R. Partington, *A History of Chemistry,* iii (1962), and others.
167. T. Webster, *Reports and Notes of Cases on Letters Patent for Inventions,* i, (1844), 80–84.
168. 32 George III, c. 72.
169. C. T. Kingzett, *The History, Products and Processes of the Alkali Trade* (1877), 73.
170. *Chemical Products,* new series xiv (1951), 179–84.
171. L. N. Vauquelin, 'Du plomb rouge de Sibérie et expériences sur le nouveau métal qu'il contient', *Journal des mines,* vi (1797), 737–60. Translated into English in *A Journal of Natural Philosophy, Chemistry and the Arts* [*Nicholson's Journal*], 1st series ii (1798), 145–46, 387–96 and 441–46.
172. L. N. Vauquelin, 'Mémoire sur la meilleure méthode pour décomposer le chrômate de fer, obtenir l'oxide de chrôme, préparer l'acide chrômique, et sur quelques combinaisons de ce dernier', *Annals de chimie,* lxx (1809), 70–94.
Vauquelin's opinion that chrome yellow was quite well known by painters in 1809 is to some extent borne out by its inclusion by James Sowerby in his book *A new elucidation of colours* (1809), 39, where he mentions that it changes green with heat, although he adds that it had not then been much used in England.
173. Clow states that an abundance of chrome ore was found in 1820 and that, for a long time, J. & J. White of Rutherglen were the only manufacturers in Scotland to make chromium compounds (*The Chemical Revolution,* 228–29). De Wild suggests that chrome yellow was first available in 1818, but does not state in which country (*The Scientific Examination of Pictures,* 59). His opinion is quoted by Gettens and Stout, *Painting Materials,* 107, and a similar remark is included in *P.P.V.,* xxxiv (1963), 470.
174. Courtauld Institute, Field/6, ff. 337v, 338, 344 and 345. See also R. D. Harley, 'Field's manuscripts: early nineteenth-century colour samples and fading tests', *Conservation,* xxiv (1979), 75–84.
175. F. Kapp, *Justus Erich Bollmann. Ein Lebensbild aus zwei Welttheilen* (Berlin, 1880), 401–04.
176. The discovery of iron chromate in Unst was made about 1817 and was reported in 1820. Royal Society of Arts, *Transactions,* xxxviii (1821), 23–26. A costed procedure for making chrome yellow appears in Berger's 'Second Experimental Notebook', f. 9.
177. *Phil. Trans.,* xlvi (1750), 584–89.
178. G. H. Bachhoffner, *Chemistry as applied to the Fine Arts* (1837), 122–23.
J. W. Mellor, *Modern Inorganic Chemistry* (1912), 642.
179. F. Stromeyer, 'New details respecting cadmium', *Annals of Philosophy,* xiv (1819), 269–74, translated from *Annalen der Physik* (Leipzig), lx.
G. H. Bachhoffner, *Chemistry as applied to the Fine Arts* (1837), 117.
180. B.M. MS. Sloane 3292, f. 3.
181. I. Bristow, 'Interior house-painting from the Restoration to the Regency', draft of unpublished D.Phil. thesis for the University of York, part 1, chapter 1, 108.
182. B.M. MS. Harley 6000, f. 8v.
183. J. Scott Taylor, *A descriptive Handbook of modern Water-Colour Pigments,* n.d. [1887], 52.

184. 'Cambogium, 3 Chestes, one runlett and a baskett. The use whereof was very much comended for a gentle purge if it were knowne in this land.' I.O.R. East India Company Court Minute Book (1613–15), 505.
185. J. Parkinson, *Theatrum Botanicum* (1640), 1575.
186. J. Smith, *The Art of Painting in Oyl* (1701), 101.
187. B.M. MS. Sloane 2052, f. 122v. De Mayerne states that pink contains chalk and that its colour is derived from *Isatis*. As *Isatis* is the Latin name for woad, which provided a blue colour, the most likely explanation is that the English names *woad* and *weld* had been confused in this instance and that de Mayerne really meant *Reseda luteola*, otherwise known as weld or dyer's weed, an annual of the mignonette family, which was used in dyeing during the medieval period until it was largely displaced by fustic.
188. B.M. MS. Additional 23080, f. 37.
189. The names listed appear in the following sources:
Pink yellow MS. Stowe 680, Peacham, John Smith.
Green pink MS. Ashmole 1494, Browne.
Light pink Goeree (English translation), Browne, Elsum, Dossie.
Brown pink Goeree, Elsum, Bardwell, Dossie.
Dutch pink Browne, *The Art of Drawing* (1731), Dossie.
Rose pink *The Art of Painting in Miniature* (1730), Dossie.
English pink *The Art of Drawing* (1731), Dossie.
Italian pink Fielding, Field.
190. Bodleian MS. Tanner 326. Quoted from E. Norgate, *Miniatura*, ed. M. Hardie (Oxford, 1919), 69–70.
191. C. Merret, *The Art of Glass* (1662), 160–61.
192. C. Merret, *The Art of Glass* (1662), 334.
193. *British Museum, Catalogue of British Drawings*, eds. E. Croft Murray and P. Hulton (1960), 440–41.
194. I. Bristow, 'Interior house-painting from the Restoration to the Regency', draft of unpublished D.Phil. thesis for the University of York, part 1, chapter 1, 108–10.
Berger, MS. 'Lake and Carmine Book', 197–98.
195. I. Bristow, *op.cit.*, part 1, chapter 1, 111.
Berger, MS. 'Lake and Carmine Book', 208.
In the formula quoted, '1 Pail Second Stuff' is not explained but it is likely that it was recycled dye from Berger's manufacture of Dutch pink, as noted in the 'Lake and Carmine Book', 127.
196. Berger, MS. 'Lake and Carmine Book', 335.
197. T. Bardwell, *The Practice of Painting* (1756), 8.
198. Patent granted 1775. Specification enrolled 1776. P.R.O. C.210/16, printed series No. 1103.
199. Extending Act, 25 G.III, c.38. Specification, P.R.O. C.210/28, printed series No. 1496.
200. A. S. MacNalty, 'Edward Bancroft, M.D., F.R.S., and the War of American Independence', *Proceedings of the Royal Society of Medicine*, xxxviii (1944–45), 7–15.
201. E. Bancroft, *Facts and Observations briefly stated, in support of an intended Application to Parliament*, n.d. [1798].
E. Bancroft, *Experimental Researches concerning the Philosophy of Permanent Colours* (1813), ii, 112–16.
202. J. Scott Taylor, *A descriptive Handbook of modern Water-Colour Pigments*, n.d. [1887], 29, 50 and 52.
203. N. S. Baer, N. Indictor and A. Joel, 'The chemistry and history of the pigment Indian yellow', *Conservation of Paintings and the Graphic Arts*, I.I.C. Lisbon Congress 1972, 401–8.
H. Kühn, 'A study of the pigments and the grounds used by Jan Vermeer', *National Gallery of Art Report and Studies in the History of Art 1968* (Washington, D.C.), 155–75.

204. Manchester Reference Library MS. 923.9. D55, Roger Dewhurst Journal, 1784–87, pp. 999–1001.
205. *Journal of the Society of Arts,* xxxii (1883–84), 16–17.
206. M. P. L. Bouvier, *Manuel des jeunes artistes* (Paris and Strasbourg, 3rd edn, 1844), 9. [A rough translation is . . . 'This yellow is not very well-known; it comes from England and I am, I believe, one of the first to have made it known on the continent. It is quite expensive; selling in London for six shillings, that is to say about six francs, per ounce. But, moreover, it is so light, and it spreads so freely, that with a single ounce one has enough for several years' use, for one employs little of it.']
207. C. L. Eastlake, *Methods and Materials of Painting of the Great Schools and Masters* (New York, 1960; reprint of 1st edn 1847), i, 446.
208. P.R.O. C.66/2357/3.
209. B.M. MS. Sloane 2052, f. 122.
210. B.M. MS. Sloane 1990, ff. 25v–26.
211. C. Merret, *The Art of Glass* (1662), 33–34, 304–06.
212. B.M. MS. Sloane 1990, f. 29v.
213. B.M. Print Room, Trade Cards, Banks Collection, Colourmen, D.2 1564 and 1565.
214. M. P. Crosland, *Historical Studies in the Language of Chemistry* (1962), 105.
215. J. Ray, *A Collection of English Words* (1674), 138.
216. R. Watson, *Chemical Essays,* iii (1782), 338–43.
217. G. Dodd, *Textile Manufactures,* v (1844), 131.
218. An example of a painting with a bright orange ground of red lead (modified with a superimposed thin layer of grey) is G. van der Eeckhout's 'Four Officers of the Amsterdam Coopers' and Wine-Rackers' Guild' (National Gallery No. 1459). Private communication from Miss Joyce Plesters.
219. Reference is made here only to the dry method of manufacture, as details of the wet method introduced in the late seventeenth century in Germany were not encountered in English sources. For an account of both methods see: R. J. Gettens, R. L. Feller and W. T. Chase, 'Vermilion and Cinnabar', *Conservation,* xvii (1972), 45–69.
R. L. Feller, 'Studies on the darkening of vermilion by light', *National Gallery of Art Report and Studies in the History of Art 1967* (Washington), 99–111.
220. D. V. Thompson, *The Materials of Medieval Painting* (1936), 102–03.
221. J. Smith, *The Art of Painting in Oyl* (1701), 19–20.
222. Tuckert, 'Mémoire sur la fabrication du sulfure de mercure sublimé (cinnabre du commerce), à Amsterdam', *Annales de chimie,* iv (1790), 25–30.
A. F. E. van Schendel, 'Manufacture of vermilion in 17th-century Amsterdam: the Pekstok papers', *Conservation,* xvii (1972), 70–82.
223. R. Ormond, 'Chinnery and his pupil, Mrs Browne', *Walpole Soc.,* xliv (1972–74), 123–214.
224. J. R. Partington, *A History of Chemistry,* ii (1961), 370–71.
225. J. R. Partington, *A History of Chemistry,* iv (1964), 85–90.
226. L. N. Vauquelin, 'Des expériences sur l'iode', *Annales de chimie,* xc (1814), 206–22 and 239–51.
227. Courtauld Institute, Field/6, f. 371. The other samples are included on ff. 362 and 384. James Ward is mentioned in connection with the pigment as well as Hobday and Sheffield. Iodide of mercury has also been identified in Turner's colour box; see N. W. Hanson, 'Some painting materials of J. M. W. Turner', *Conservation,* i (1954), 162–73.
228. J. Scott Taylor, *A descriptive Handbook of modern Water-Colour Pigments,* n.d. [1887], 38.
Berger, MS. 'Second Experimental Notebook', ff. 23 and 85, dated 1830 and 1833.
229. *Annales de chimie,* lxx (1809), 90–91.
230. J. Bodams, 'On a scarlet sub-chromate of lead, and its application to painting and calico printing', *Annals of Philosophy,* new series ix (1825), 303–5.

231. Kermes (associated with grain, a colour mentioned in medieval sources) is not discussed in any detail here, for, although the European kermes insect *Coccus ilicis* was discussed during the seventeenth century in *Philosophical Transactions,* its absence from documents concerned specifically with painting suggests that the improved availability of Indian lac and Latin American cochineal in England by the beginning of the seventeenth century had put an end to the use of the European product for painting purposes.
232. *Shellac* (Angelo Brothers Ltd, Calcutta, 1965), 14.
233. W. F. Leggett, *Ancient and Medieval Dyes* (Brooklyn, 1944), 70.
234. D. V. Thompson, *The Materials of Medieval Painting* (1936), 111.
235.

Limming (1573)	sinapor lake, sinapor topias.
V. & A. MS. 86.EE.69	sinoper lake, sinoper tappes.
Bodleian MS. Ashmole 1494	sinaper lacke, sinaper tape.
B.M. MS. Sloane 3292	sinaper lacke.
B.M. MS. Sloane 1394	sinaper.
B.M. MS. Stowe 680	sinoper lake, sinoper topps.
Peacham	sinaper lake, sinaper tops.

236. H. Peacham, *The Gentlemans Exercise* (1612), 87.
237. J. Parkinson, *Theatrum Botanicum* (1640), 1588.
238. W. F. Leggett, *Ancient and Medieval Dyes* (Brooklyn, 1944), 91.
239. B.M. MS. Sloane 6284, f. 109.
240. B.M. MS. Stowe 746, f. 57.
241. While there is a remote possibility that the English lake referred to was English gum hedera, which is mentioned by Peacham and others as a constituent of a water-colour medium known as gum lake, it is far more likely that the lake was made in England from cochineal, either imported in the normal way or seized from Spanish ships which were attacked in the Atlantic on their way home from the Spanish American colonies. Reports on the cargo carried by Spanish and British ships from America were sometimes sent by the Venetian ambassador to the council at Venice. Cochineal was obviously a valuable commodity worth listing; on one occasion it was described, with pieces of eight, as forming a valuable shipment worth £100,000. *Calendar of State Papers, Venetian,* xxvi, 207 and 214.
242. W. F. Leggett, *Ancient and Medieval Dyes* (Brooklyn, 1944), 86.
243. *Phil. Trans.,* xvii (1693), 502–03.
244. W. F. Leggett, *Ancient and Medieval Dyes* (Brooklyn, 1944), 83–86.
245. B.M. MS. Sloane 1394, f. 139.
246. J. Smith, *The Art of Painting in Oyl* (1701), 100.
247. Berger, MS. 'Lake and Carmine Book', 6.
248. I. Bristow, 'Interior house-painting from the Restoration to the Regency', draft of unpublished D.Phil. thesis for the University of York, part 1, chapter 1, 143.
249. Van Dyck, 'Portrait of Lady Elizabeth Thimbleby and Dorothy, Viscountess Andover' (National Gallery No. 6437). Private communication from Miss Joyce Plesters. For information on the methods used for identification of the raw materials used for the dye in red lake pigments, see J. Kirby, 'A Spectrophotometric Method for the Identification of Lake Pigment Dyestuffs', *National Gallery Technical Bulletin,* i (1977), 35–44.
250. T. Bardwell, *The Practice of Painting* (1756), 8.
251. B.M. MS. Sloane 2052, f. 29.
252. D. V. Thompson, *The Materials of Medieval Painting* (1936), 121.
253. J. Gerard, *The Herball* (1633), 1120.
254. W. F. Leggett, *Ancient and Medieval Dyes* (Brooklyn, 1944), 7–12.
255. P.R.O. C.66/2338/20.
256. P.R.O. SP.16/315/141 and SP.16/321/19.
257. P.R.O. SP.16/323/54 (land for madder growing).
P.R.O. C.66/2749/2 (indenture).

258. E. Bancroft, *Philosophy of Permanent Colours* (1813), ii, 223. The French equivalent for each of the special terms is listed in the same place:

Mull madder	*garance courte.*
Gemeen madder	*garance mi-robée.*
Crop madder	*garance robée* or *garance grappe.*

259. D. Hudson and K. W. Luckhurst, *The Royal Society of Arts* (1954), 89–90.
260. W. F. Leggett, *Ancient and Medieval Dyes* (Brooklyn, 1944), 15.
261. C. Merret, *The Art of Glass* (1662), 178.
262. Royal Society of Arts, *Transactions,* xxii (1804), 141–61.
263. Royal Society of Arts, Minutes of Committee of Chemistry and Mechanics (1814–15), 225, 238 and 277; and (1815–16), 94–96 and 100.
264. Royal Society of Arts, *Transactions,* xxxiv (1817), 87–94.
265. M. J. F. L. Mérimée, *The Art of Painting in Oil* (1839), 130–50.
266. *The Gentleman's Magazine,* new series xlii (1854), 524–25.
267. Courtauld Institute, Field/2, unnumbered folio following title page and preceding f. 1. The relevant passage and some notes on an experiment in making purple madder are quoted in R. D. Harley, 'Field's manuscripts: early nineteenth-century colour samples and fading tests', *Conservation,* xxiv (1979), 75–84.
268. A transcript of Thirtle's manuscript treatise is included in M. Allthorpe-Guyton, *John Thirtle, 1777–1839: Drawings in Norwich Castle Museum* (exhibition catalogue published by Norfolk Museums Service, Norwich, 1977).
269. B.M. MS. Sloane 122, f. 70v, and V. & A. MS. 86.EE.69, f. 62.
270. B.M. MS. Sloane 3292, f. 6.
271. B.M. MS. Sloane 2052, f. 62.
272. W. Goeree, *The Art of Limning* (1674), 7.
273. The difference between lake-making with brasil and mordanting a white pigment with the dye is discussed by D. V. Thompson, *The Materials of Medieval Painting* (1936), 118–19.
274. B.M. MS. Harley 6000, f. 17.
275. J. W. Alston, *Hints to young Practitioners in the Study of Landscape Painting,* n.d. [1804], 62.
276. N. W. Hanson, 'Some painting materials of J. M. W. Turner', *Conservation,* i (1954), 164.
277. B.M. MS. Harley 6000, f. 3.
278. B.M. MS. Harley 6000, f. 3.
279. R. J. Gettens and G. L. Stout, *Painting Materials* (New York, 1942), 168.
280. B.M. MS. Harley 6000, f. 3.
281. J. Parkinson, *Theatrum Botanicum* (1640), 1573.
282. B.M. MS. Sloane 2052, f. 95.
283. Anon., *Practical Treatise on Painting in Oil-Colours* (1795), 215.
284. W. G. Constable, *Richard Wilson* (1953), 114.
285. Wilkie praised asphaltum in a letter in 1823: 'The pot of asphaltum, if you can get it mixed in such a way as not to run, will give a fine rich surface to what you do. It is the most gorgeous of colours.' Quoted from W. T. Whitley, *Art in England 1821–37* (Cambridge, 1930), 44.
286. J. Sanderson, *The Travels of John Sanderson in the Levant 1584–1602,* ed. W. Foster (Hakluyt Society, London, 1931), 44–45.
287. A. Browne, *Ars Pictoria* (1675), Appendix, p. 3.
288. Anon., *A Compendium of Colors,* n.d. [1797], 221.
289. Courtauld Institute, Field/6, f. 374.
290. Anon., *A Compendium of Colors,* n.d. [1797], 222.
291. J. Bryant, *Bryant's Treatise on the Use of Indian Ink and Colours,* n.d. [1808].
292. A. Ure, *A Dictionary of Arts, Manufactures, and Mines* (1839), 1098.
293. 'Small-Coal is prepar'd from the Spray, and Brush Wood, stripp'd off from the Branches of Coppice Wood; sometimes bound in Bavins for that purpose,

and sometimes prepar'd without binding. The Wood they dispose on a level Floor, and setting a Portion of it on fire, throw on more and more, as fast as it kindles; whence arises a sudden blaze, till all be burnt that was near the Place. As soon as all the Wood is thrown on, they cast Water on the Heap, from a large Dish, or Scoop; and thus keep plying the Heap of glowing Coals, which stops the Fury of the Fire, while with a Rake they spread it open, and turn it with Shovels till no more Fire appears. When cold, they are put up into Sacks for use.' Quoted from *Cyclopaedia,* ed. E. Chambers (1728), *s.v.* Coal.

294. B.M. MS. Harley 6376, p. 108.
295. B.M. MS. Sloane 2052, f. 18.
296. H. Peacham, *The Art of Drawing with the Pen* (1606), 60.
297. B.M. MS. Sloane 2052, f. 6 (instructions mentioning salt) and f. 93 (instructions attributed to Mytens).
298. H. Peacham, *The Art of Drawing with the Pen* (1606), 60.
299. *Encyclopédie,* eds. D. Diderot and J. le R. d'Alembert (Paris and Neuchâtel, 1754–65), xxiii.
300. B.M. MS. Sloane 1990, f. 60v.
301. H. Peacham, *The Art of Drawing with the Pen* (1606), 56.
302. B.M. MS. Harley 6000, f. 16v.
303. B.M. MS. Sloane 2052, f. 11v
A. Browne, *Ars Pictoria* (1675), 77.
C.B. [C. Boutet], *Traité de mignature* (1674), 14.
See J. A. van de Graaf, *Burlington Magazine,* civ (1962), 471–75, for evidence of lead white and chalk priming found during the scientific examination of Dutch paintings.
304. B.M. MS. Harley 6000, f. 1v.
305. B.M. MS. Sloane 2052, f. 88.
306. Anon., *The Art of Painting in Miniature* (1729), 12.
307. B.M. MS. Sloane 122, f. 92v.
308. D. V. Thompson, *The Materials of Medieval Painting* (1936), 92–93.
See also R. J. Gettens, H. Kühn and W. T. Chase, 'Lead white', *Conservation,* xii (1967), 125–39.
309. B.M. MS. Harley 6000, f. 17.
310. B.M. MS. Sloane 1990, f. 26.
B.M. MS. Sloane 2052, f. 6.
B.M. MS. Stowe 680, f. 132.
311. P.R.O. C.66/2271/32, printed series Nos. 22 and 22*.
312. P. Vernatti, 'A Relation on the Making of Ceruss', *Phil. Trans.,* xii (1678), 935–36.
313. R. Watson, *Chemical Essays,* iii (1782), 361.
314. An explanation of chemical reactions in the formation of stack white lead is included, with chemical formulae, in A. and N. L. Clow, *The Chemical Revolution* (1952), 382.
315. T. Birch, *The History of the Royal Society* (1756–57), iii, 517.
316. *Repertory of Patents of Invention,* v (1796), 249.
317. B.M. MS. Harley 6376, p. 42.
318. *Repertory of Patents of Invention,* v (1796), 138.
319. Royal Society of Arts, *Transactions,* xxii (1804), 260–74, and MS. Minutes of Committees, Mechanics, 1803–4, 237–38.
320. J. Smith, *The Art of Painting in Oyl* (1701), 16.
321. B.M. MS. Harley 6000, f. 13v.
322. Anon., *The Art of Painting in Miniature* (1729), 95.
323. Anon., *The Art of Drawing* (1731), 23.
324. Anon., *The Art of Drawing* (1731), 23 (MS. additions in V. & A. copy, press mark 41. DD.8).
325. B.M. MS. Sloane 288, f. 130v.
326. B.M. MS. Sloane 2052, f. 10v.

327. B.M. MS. Sloane 1990, f. 54 and f. 86.
328. R. Boyle, *Experiments and Considerations touching Colours* (1664), 342 and 358.
329. L. B. Guyton de Morveau, 'Recherches pour perfectionner la préparation des couleurs employées dans la peinture', *Nouveaux mémoires de l'Académie de Dijon* (Dijon), i (1782), 1–24.
330. J. R. Partington, *A History of Chemistry,* ii (1961), 51.
331. B.M. MS. Sloane 2052, f. 92.
332. R. Watson, *Chemical Essays,* iii (5th edn 1798), 364.
C. Mackenzie, *1000 Processes* (1825), 156.
333. C. de Massoul, *A Treatise on the Art of Painting* (1797), 135.
334. H. Kühn, 'Une boîte à couleurs française du debout du XIXe siècle', *P.P.V.,* xxxix (1963), 477.
335. G. Agricola, *De re metallica,* eds. H. C. and L. H. Hoover (1912), 115 note.
336. A. Ure, *A Dictionary of Arts, Manufactures, and Mines* (1839), 1297–98.
337. S. Parkes, *Chemical Essays,* ii (1815), 195.
338. E. Brown, *A Brief Account of some Travels in divers Parts of Europe* (2nd edn 1685), 184–85.
339. C. Merret, *The Art of Glass* (1662), 300.
340. P. Shaw, *Chemical Lectures* (1733), 282.
341. *Nouveaux mémoires de l'Académie de Dijon* (Dijon), i (1782), 1–24.
342. Anon., *The Artist's Repository,* n.d. [1784/6], ii, 75.
343. P.R.O. C.210/47/1 and C.210/51/15. Printed series Nos. 1996 and 2094.
344. J. C. Ibbetson, *An Accidence or Gamut of Painting* (1803), 15.
345. W. Winsor and H. C. Newton, *Remarks on the White Pigments used by Water-Colour Painters* (1837).
346. Anon., *Practical Treatise on Painting in Oil-Colours* (1795), 25.
The Artist (1810), ii, 199–205.
347. *The Artist* (1810), ii, 205.
348. G. Field, *Chromatography* (1835), xi.
349. Q. Bell, *The Schools of Design* (1963).
350. P.R.O. C.66/1682/49.
351. P.R.O. SP.14/22/30.
352. P.R.O. SP.14/22/30.
353. *House of Commons Journals,* i (1547–1628), 317.
354. P.R.O. C.66/1751/10.
355. P.R.O. SP.14/72/84.
356. P.R.O. SP.14/72/89.
357. *Publications of the Huguenot Society of London* (Aberdeen), x (1900–8), part 3, 227.
358. P.R.O. SP.14/97/90–92.
359. P.R.O. C.66/2178/13 and SP/14/141/242.
360. P.R.O. SP.14/145/18. Suits in such form were heard in the Court of Requests, but there are no documents relating to Abraham Baker in the relevant year.
361. P.R.O. E.190/34.
362. I.O.R. East India Company Court Minute Book, xiv, 294.
363. P.R.O. C.66/2644/25.
364. P.R.O. C.66/2357/3.
365. I.O.R. East India Company Court Minute Book, iv, 241.
366. I.O.R. East India Company Court Minute Book, xv, 67.
367. I.O.R. East India Company Court Minute Book, xv, 114.
368. I.O.R. East India Company Court Minute Book, xv, 118.
369. I.O.R. East India Company Court Minute Book, xvi, 21–2.
370. I.O.R. East India Company Court Minute Book, xvii, 334.

Index

Page numbers in bold type indicate the main references to the topic concerned.

SMALT: SEE PAGE. 53.

SMALTS BOUGHT BY JAMES WHATMAN II THE PAPER MAKER DURING 1782:

Three grades of Smalts appear in the account books:

OC. of which Whatman bought 379 lbs, at 7¾ d per lb.

FFFE. of which he bought only two consignments totalling 3.786 lbs, at 20d to 22d a lb.

FFE. on which he bought 41.281. lbs over a period of eight years, generally at 15½ d to 16 d per lb — but in 1782 at 12¾ d a lb.

∴ The annual expenditure on smalts was about £370

The smalts were presumably employed in the furtherance of his (Whatman II) experiments to correct the yellow tinge of his paper: But he may also have been making some 'Azure' papers.

see James Whatman, father & Son by Thomas Balston 1957. Reprint Garland Books. NY & LONDON 1979